# Mixed Medicines

# Mixed Medicines

## Health and Culture in French Colonial Cambodia

SOKHIENG AU

The University of Chicago Press
Chicago and London

**Sokhieng Au** is an independent scholar specializing in the history of medicine and Southeast Asian studies. Her interdisciplinary research focuses on a range of topics including colonial medicine; cultures of disease in Southeast Asia; medicine and gender; and, most recently, international public health.

The University of Chicago Press, Chicago 60637
The University of Chicago Press, Ltd., London

Printed in the United States of America

20 19 18 17 16 15 14 13 12 11 1 2 3 4 5

ISBN-13: 978-0-226-03163-7 (cloth)
ISBN-13: 978-0-226-03164-4 (paper)
ISBN-10: 0-226-03163-2 (cloth)
ISBN-10: 0-226-03164-0 (paper)

Au, Sokhieng.
Mixed medicines : health and culture in French colonial Cambodia / Sokhieng Au.
p. cm.
Includes bibliographical references and index.
ISBN-13: 978-0-226-03163-7 (cloth : alk. paper)
ISBN-10: 0-226-03163-2 (cloth : alk. paper)
ISBN-13: 978-0-226-03164-4 (pbk. : alk. paper)
ISBN-10: 0-226-03164-0 (pbk. : alk. paper) 1. Public health—Cambodia—History. 2. Cambodia—History—1863–1953. I. Title.
RA407.5.C16 A93 2011
362.10959622
2010029463

♾ The paper used in this publication meets the minimum requirements of the American National Standard for Information Sciences—Permanence of Paper for Printed Library Materials, ANSI Z39.48-1992.

CONTENTS

# ACKNOWLEDGMENTS

As a refugee from Khmer Rouge Cambodia, I spent a good portion of my early adult life avoiding the topic of Cambodian history, because it seemed in both scholarly and popular media to be saturated with sensationalistic accounts of the horrors of war and self-inflicted violence. I had no interest in identifying with either victim or perpetrator. However, during my early years in the history PhD program at the University of California, Berkeley, I came to realize that huge and fascinating swaths of Cambodian history had yet to be explored by Western academic scholarship. Ultimately, the research for this book allowed me to combine my new interest in the rich and neglected histories of Cambodia with my previous interests in the history of science and medicine and in disciplines such as medical anthropology and medical sociology. Along this path, I met many amazing and wonderful people.

In Cambodia, the staff of the National Archives provided me not only with research assistance but also with friendship, encouragement, and a steady supply of snacks and gossip. I will always be glad that I had the opportunity to meet my sister at the archives, Y Dari (also known as chief administrator of the Technical Bureau), and "Auntie" Hou Rin (chief administrator of the Repository Bureau). I am thankful as well to Chhem Neang, Chun Lim, and Mam Chean. Additionally, I will forever be indebted to Peter Arfanis and Greg Müller, who were then working with the National Archive staff.

Research in France would not have been possible without funding from the Peder Sather Chairholder Fellowship, the Pacific Rim Research Grant, the Erhman Traveling Fellowship, and the Graduate Opportunity Program Fellowship. For my research in France, I thank the staff of the Centre des Archives d'Outre-Mer, the Bibliothèque National, and the École Française d'Extrême Orient. In particular, the kind and helpful staff at the archives of the Pasteur Institute in Paris made research there a pleasure. My research in Cambodia was supported by the Luce Foundation Dissertation Research Fellowship, administered by the Center for Khmer Studies. The center and its then director, David Chandler, were immensely supportive during my field research in Cambodia. During the revision of this book I was supported by a fellowship from the Science in Human Culture Program at Northwestern University. I also owe tremendous thanks to the program's director, Ken Alder, for his mentorship, insightful advice, and careful editing.

I have received assistance and advice from many inspiring colleagues and mentors, including Shawn Morton (who started me on this path), Jack E. Lesch, Cathryn Carson, Peter Zinoman, Jeffrey Hadler, Penny Edwards, Francesca Bordogna, Anne Hansen, Mark Bradley, Judy Ledgerwood, Laurence Monnais, Ashley Thompson, the Southeast Asian studies and history of science communities at UC Berkeley, and the Science in Human Culture community at Northwestern University. I am also grateful to my peers in the field and in the classroom who have shared their ideas and research with me, including Navin Moul, Jennifer Foley, Gerard Sasges, Soizick Crochet, Rethy Chem, Anne Yvonne Guillou, Sandro Wadman, Martina Nguyen, Tony Hazard, Laura Stark, Alistair Sponsel, Asia Nguyen, Annick Guénel, and Matthieu Geurin.

Of course, I must thank my loving partner William Etienne, my sisters and their families, and my mother. I also have to thank my father, deceased these many years. He continues to remind me that the memories we keep and the stories we share about those who are gone matter, as even in their absence they affect us profoundly.

# ABBREVIATIONS

## ARCHIVAL SOURCES

AIP — Archives of the Institut Pasteur

CAOM — Centre des Archives d'Outre Mer (French colonial archives, Aix-en-Provence)

CAOM BIB — Library collections

CAOM INDO — Indochina collections

CAOM INDO RSC — Résidence Supérieur du Cambodge collections

CAOM INDO RSTNF — Résidence Supérieur du Tonkin collections

EFEO — École Française d'Extrême Orient

NAC — National Archives of Cambodia

NAC RSC — National Archives of Cambodia, Collection of the Résidence Supérieure du Cambodge

## GOVERNMENT AND MEDICAL SERVICES

AM — Assistance Médicale (its full title was Assistance Médicale Indigène)

DLS — Directeur local de la santé (local director of health)

GGI — Gouvernement Général de l'Indochine (General Government of Indochina)

PI — Pasteur Institute

RSC — Résident supérieur du Cambodge

SPE — Société de Protection de l'Enfance (Society for the Protection of Infancy)

SPMI — Société de Protection Maternelle et Infantile au Cambodge (Society for Maternal and Infant Protection of Cambodia)

# INTRODUCTION

This book offers more than a history of Southeast Asia or a history of medicine, although its proximate subject lies at their intersection: the attempts by the French colonial health services to expand Western medicine in Cambodia. French colonial involvement in Cambodia dates from 1863 to 1954; this study, however, focuses largely on the period of 1907 to 1940, the years which saw the greatest proliferation of hospitals, clinics, and laboratories; the most substantial growth in medical practitioners and researchers; and increased discussion, legislation, and education in health-related fields. This book has three broad aims: to contribute to the burgeoning field of historical studies of globalization, in part by showing the connections between history of medicine in the colonial context and the evolution of medicine in Europe; to highlight the vital role of colonial peripheries in shaping the colonial experience; and to provide a detailed social history of Cambodia that contributes meaningfully to Southeast Asian history.

The history of medicine as a field has increasingly focused on the complexity of the development of "Western" science in non-Western contexts. This book belongs to a newer movement in the discipline that seeks to historicize the globalization of medicine. This historicization—borne of the essentially economic concerns of contemporary studies in globalization—is now beginning to encompass noneconomic aspects of this process, such as local cultures. *Mixed Medicines* approaches this problem in an innovative manner, paying attention to a phenomenon that I term cultural insolubility. Medicine in Cambodia essentially existed in a plural state. The gap between Western and indigenous medicine grew during the colonial era. What drove this paradoxical phenomenon? Indigenous people rejected French medicine not only because of the dynamics of colonialism, but also because of the dynamics of cultural distinction. In the current environment of globalization, issues of integration and acculturation are paramount. Local actors are either applauded or reviled for their "resistance" to the trends of a globalizing world. Little attention is paid to the ways in which different sorts of networks and cultural systems cannot be smoothly integrated because they would lose internal coherence. While globalization studies frame "resistance to integration" in political and economic terms, other sorts of network theorizing, notably from sociological studies examining end-user transformations of technology, have revealed how end-users transform the meaning of both professional practices and material objects.[1] That which cannot be made culturally meaningful is generally rejected; this is true in both the French and Khmer contexts. However, while human agency is important, we must be mindful of the tendency for complex social systems (or, shall we say, cultures) to maintain their internal coherence. The Khmer and French actors within these systems are not always explicitly political (meaning that they are not always attempting to influence interpersonal power relations) when they operate. The systemic constraints on actors are often neglected in historiographical reconstruction because we focus on change and view stabilizers either as active resistance to change or obstacles to change. The history that follows explicates stasis as much as transformation.

Again, this is not a study of Cambodian medicine versus Western medicine. There can be little dispute today that the therapeutic efficacy of Western medicine for acute (not chronic) diseases is far greater than that of traditional Cambodian medicine. But there are other aspects of healing and medicine where the metrics are quite muddled. The meanings ascribed to Western medicine—the various controversies and misunderstandings in the story that follows—give us an understanding of a particular worldview, of an experience that has passed. I try to avoid judging too harshly the men and women whose triumphs and failures (and they were, for the French medical establishment, largely failures) follow. My sympathies, if they must be declared, are with all the actors attempting to comprehend and negotiate differences to a beneficial end. This is a history of European doctors far afield and of villagers intruded upon in their homes by aliens. Obviously, two people can experience the same event in the same place at the same moment and understand it in completely different ways. They live in two different thought worlds. Culture, declared Geertz, is the web of significance within which we are suspended and through which we give meaning to the events of our lives.[2] It is a whole system that situates human experience. The colonial moment is so interesting for science studies and Khmer history because of the multiplicity of thought worlds and cultural values that operate and try to define events as they occur. These cultural values frequently do not blend in the moment of the event.

In the stories that follow, specific vignettes are offered as indications of, but not as representative of, the whole. I seek analytical parity, but not in individual concepts or practices. In other words, it is not productive to examine a specific idea of the body in French culture against a "comparable" idea of the body in Cambodian culture to understand why conflict occurred. But it is productive to trace the disparate valences of meaning arising from the conflict. A Western medical construct of "the body" comes in conflict with Cambodian political identities, and Cambodian ritual practices conflict with French civil values. Cultural insolubility is a way to speak of those subalterns whose presence we seek through alternative sources in history. These alternate, often non-textual, sources give us the ghost of their presence. Like the shadows cast in Plato's allegorical cave, they provide us some sense of another reality that we cannot see directly. The stories herein do not squeeze too much significance from these shadows, but they do acknowledge that that the absence of these subalterns tells us something. Human beings tend in general to be repelled by what they do not understand and drawn to what is familiar to them. If we cannot understand something, we simplify it to a level at which we can understand it. I refer again not only to Cambodians but also to the French. Cultural insolubility can limit the diffusion of one culture through others, but this is driven in part by a tendency for people to stay within the enclaves of their own cultural thought worlds.

This study remains mindful of medical scholarship on Europe when contextualizing European medical practitioners in colonial contexts. The relationships between colonizer and colonized, doctor and patient, and researcher and subject can appear deceptively similar. The only way to reconstitute the motives, the experiences, and particularly the differences in medical practice between the metropolitan and colonial domains is by considering the wide-ranging factors that drove the transformation of medicine globally at the turn of the twentieth century. Also, the peculiar context of the Khmer professionals in the French medical corps must be taken into account. In her recent study of Khmer nationalism, Penny Edwards has described the Khmer elite within the colonial services as "cross-cultural interlocutors."[3] Similarly, Anne Hansen has observed that Buddhist modernists in Cambodia during this time were liminal characters who represented "Southeast Asian religious engagement with modernity."[4] The Khmer medical professionals were elite, but they were of the new secular elite. Within a Weberian scheme, their authority was bureaucratically rather than traditionally or charismatically based.[5] However, bureaucratic authority was not cultural authority. The colonial moment is interesting because of the wide divergence in the sources from which respect, authority, and obedience derive. A healer had morality/merit/power derived from *dhamma*, while an M.D. had facts, practices, and certifications derived from study. The Cambodian patient was not quick to validate this new sort of expertise. Also, the first generation of Cambodian Western doctors, nurses, and midwives had careers that depended on two different concepts of social organization. Their careers and personal lives serve as proof of insolubility as well as of hybridization or modernization. The stories within this book emphasize that the evolution of medicine in both the European and indigenous contexts is shaped by intimate local engagements as well as global changes in medicine, and specific social developments in Cambodia and metropolitan France.

The increasing reach of Western medical services in the early twentieth century, as well as the expansion of the colonial state more generally, laid the groundwork for profound changes in Cambodian society. However, colonial medical services were not entering a health care vacuum. Indigenous people of Khmer, Cham, Vietnamese, and other ethnicities who made up the population of Cambodia observed a wide variety of medical practices in the precolonial period. As the French health services encroached beyond colonial enclaves, the technologies they employed along with their cultural associations had to negotiate with existing cultural practices. Ultimately, the Western contractual relation between doctor and patient never effectively replaced the preexisting modes of social relations in the village. The Western doctor was not and did not become the equivalent of a Cambodian healer. Nonetheless, changes did occur. Both European and indigenous practices were transformed in unexpected ways during this period of interaction. However, the two systems of health neither effectively hybridized nor displaced one another; rather, they had distinct narratives that at some times overlapped and at others developed in parallel. This is not to say that there are neat lines between Cambodian and French medicine. They are not binary opposites. In fact, binaries are precisely what I would like to avoid. Too often, histories of Western medicine in the colonies, sympathetic to the "opposition" from the subaltern populations, seek and define that opposition in terms of contrasting medicines. All medical practice and theory that cannot mirror or resist its Western counterpart is excluded from such a dialectical model of medical exchange. I do not attempt in this study to do justice to the expansive domain of Cambodian healing practices and theory. Like the histories I criticize, I am confining myself to those bits of medicine that abut French medical practices. However, I am also including a more nuanced and expansive analysis of Cambodian social practices and concepts that deeply affect inception of French medicine. The comparison being made is not between French and Cambodian medicine; it is between concepts of the body, of politics, and of social relations along the fault line of French medical interventions.

While this is a study of Western medicine in colonial Cambodia, it does not assume that Western medicine or colonialism is the engine of the story. One obvious but perhaps radical premise of my analysis is that oftentimes Frenchmen were not providing what the indigenous populations considered medicine. This becomes more apparent when one considers that healing systems are just that: systems. The cultural context is part and parcel of how these systems work. The stories in this study portray how medical processes and technologies attain (or retain) cultural meaning. Recent work in the sociology of science has highlighted the importance of networks in the development of technological systems. This literature emphasizes the creative action of end-users who help construct the eventual meanings and applications of technologies. These studies have also highlighted the importance of user resistance, non-use or limited availability of technologies, and the role of the state in configuring technologies that involve both individual needs and state concerns. While developed largely in the context of the industrialized West, these concepts are also central to medical science as it develops in the colony.[6] Such a framework is thus useful in allowing us to understand indigenous "networks" of medical technology with parity to more standard "technological networks" within science and technology studies.

As a Southeast Asian area studies monograph, this book brings Cambodia to the fore. Often marginalized in modern Southeast Asian histories due to its provincialized relationship to Vietnam, the Cambodian case is so frequently folded into "Indochinese" history that it is a source of frustration to scholars and instructors of mainland Southeast Asia in the modern era. The concepts of centers and peripheries are key to my analysis. I am calling into question the very labeling of the periphery as such. While much has been written on frontiers and boundary societies in the creation of nation-states and national identities, little has been written of the importance of the experience of the periphery in these transformations. For most French colonial officials, and even more for colonial subjects, the experience of the center is not representative of their own historical experience. Yet centers are taken to represent historical experience in most colonial histories. In the specific case of Cambodia, almost no subjects shared the language or culture of the "local" colonial metropole of Saigon, and even fewer that of distant Paris.

This study pays attention to the specific sociocultural, religious, political, geographical, and of course historical context of Cambodia, all of which is vastly different from that of Vietnam. Most social histories of "Indochina" are in fact social histories of the three states of what is now present-day Vietnam. Indochina is a French construct. It is a valid category to use when examining French conceptions, such as how the colonial government viewed its subject populations. Later in the colonial period, as Christopher Goscha has pointed out, the Vietnamese began to conceive of themselves in similar terms.[7] However, for most of the colonial period in Vietnam, and always in Cambodia, subject populations did not conceive of themselves as Indochinese. The dynamic between different ethnic groups, although little studied, significantly shaped indigenous response to colonial policies in Cambodia. Ironically, the historian's tendency to subsume Cambodia within larger Indochina studies is a historical artifact of the colonial era.

Thus, this history disentangles the relationship between France and Indochina from the relationship between France and what it saw as its minor frontier colony of Cambodia. Cambodia had two metropoles: the smaller metropole of Saigon (and, for medicine, Hanoi) and the larger one of Paris. French administrators had a different relationship to this "backwater" than to the major states of present-day Vietnam. The dynamic between Cambodia and France is further distinguishable from, if entangled with, the one between Indochina and France. Again, these lesser political networks impacted policies in Cambodia—how they were enacted, carried out, and received. While it examines "Indochina" when appropriate, this study is sensitive to the specificity of Cambodia in its relations with the colonial government.

Many existing colonial medical literatures fall into three general categories: medicine as tool of empire, colonial medicine as diffusion of or supplement to the development of European medicine, and colonial medicine and its demographic impacts. In some of the best theoretically inclined histories of colonial medicine, Western medicine is generalized as a tool of empire.[8] Such scholarship links science to empire, and thus to larger literatures in imperialism and colonialism. Medicine colludes with empire—hence, colonial efforts to expand Western medicine also represent an expansion of exploitation. In contrast, and sometimes within the same scholarship, the lack of medical commitment is read as the lack of true concern for the indigenous population. The failure to offer medical care is proof of exploitation. Paradoxically, the colonist exploits both by imposing Western medicine and by withholding it. While condemning the colonist's project of discursively creating a subject native population through medical discourse and technology while simultaneously silencing the indigenous voice, such an approach itself can also fail to characterize the living experiences and active response of these same populations.[9] More recent literature recognizes that such framings neglect other important processes at work. For example, some scholarship—strongly informed by the subaltern theoretical program coming out of India—recognizes the role of indigenous choice in guiding how colonial health care is implemented.[10]

Another category of colonial medical history concerns itself with the development of European medicine in the colony, often accepting that the colonial experience shaped developments in Europe without explicitly drawing the links. Such histories deemphasize the theoretical implications of medicine within the colonial context, while providing a much-needed if less glamorous descriptive understanding of European medicine in the colony.[11] While serving an invaluable role in laying the groundwork for an understanding of colonial scientific institutions, they often isolate the medical project as a diffusion of Western medicine into the colony. Colonial medicine is also used as part of historical demographic studies, either to rebut the medicine-as-tool-of-empire scholarship or to prove the demographic devastation of the colonial project.[12] In each case, the indigenous populations exist as little more than data assessing the impact of foreign medical measures, serving an essentially passive role within both descriptive and demographic histories.

Indigenous populations are neither only haplessly exploited guinea pigs nor proof of colonialism's positive effects. Similarly, the extension of Western medicine cannot be reduced to either "tool of empire" or its converse, "token of benevolence." Ultimately, both extremes must be abandoned. This study does look at the changes to individual and population health, but its concerns lie much more with cultural interactions. It does not ask whether Western medicine is good or bad; it seeks to understand the processes whereby vastly different cultures interacted and attempted to negotiate with and understand each other, the epistemological wrangling that resulted, and the wider social implications of the interface between Western medicine and Cambodian society.

## The Framework of This Study

The stories that follow provide both a wide view of global trends in the history of medicine and a sharp focus on a social history of various interacting characters in a specific historical milieu. Chapter 1 introduces the setting for the development of Western medicine in French colonial Cambodia. It details the historical context of French intervention in the region, as well as the trends in Western, and more specifically French, medicine in the nineteenth and early twentieth centuries. It provides a sketch of the major professional, technical, and conceptual changes in medicine and biomedical research, along with Khmer medical practices and philosophies. Because Khmer medicine—indeed, all medicine—is deeply embedded in the social and metaphysical world, chapter 1 also contextualizes the social milieu into which French colonizers were venturing.

The three chapters that follow, while organized around specific themes, also correspond roughly with the three key phases in the French medical programs: mass vaccination programs (chapter 2); the formation of the colonial health service, the Assistance Médicale (chapter 3); and the social hygiene movement (chapter 4). Chapter 2 situates the French medical practitioners and researchers in Indochina. To provide an understanding of the French colonial government's commitment to aspects of science and medicine that appeared deeply flawed, it outlines the structure of the three wings of the French medical service and specific instances of vanguard medical research. Focusing on controversial incidents surrounding vaccine research and the Pasteur Institutes in Indochina, chapter 2 reveals the links between science, national pride, political exigency, and imperial ambitions.

Chapters 3 and 4 describe the operations of the French medical establishment in Cambodia. Chapter 3 considers the infrastructure of the new medical service for the indigenous population, the Assistance Médicale, and its context of operation. The socioeconomic and geographic milieu in Cambodia demanded a certain flexibility from colonial administrators. The two systems of medicine, Khmer and French, operated in very different social spheres. If the colonial government wanted medical influence to extend beyond epidemic control, it had to become more flexible. While attempting to recruit indigenous staff to expand the Assistance Médicale's personnel base, it also endeavored to accommodate the customs and concerns of indigenous groups in hospital programs and the regulation of traditional doctors and medicinal sales. The histories of staff recruitment, the hospital system, and pharmaceutical control in the country illustrate the frictions that arose from the radically different Khmer and French definitions of sickness, disease, and healing.

Chapter 4 focuses on public health, mass education, and the colonial shift towards population health and away from individual health. The increased interest in social medicine and eugenics in 1930s France led to a new drive in public health services in its colonies. In Cambodia, the colonial government mimicked various techniques from the metropole in an effort to expand the public health sector. More often than not, it was stymied by an unwilling population. Further, the difficult physical terrain, climate, and lack of physical infrastructure limited mobility through the country. Certain Khmers may have participated in programs such as the "better baby contest," but most stayed away, opting for their own preexisting "social medicine." The contrast between the two types of "social medicine"—which in the Khmer context was the traditional form of medicine performed by individuals with key roles in religious and social life, and in the French context was the institution of contracted forms of health surveillance, public education, and pharmaceutical dispensation—reveals the disjuncture of French intent and Khmer perception.

Chapter 5 examines the contradictions in the colonial government's approach to indigenous women's health. Women largely avoided all aspects of French medicine. The colonial government's interest in women's health came late, and was driven primarily by interests in population and military health. The medical service undertook several different approaches in its attempt to bring native women under its medical purview. The most significant of these efforts were three midwife training programs that took place from the turn of the century to the 1930s. The evolution and failure of these programs reveal the tensions that arose when the traditional roles of the woman in the family and in the public sphere were transformed under the French medical service.

Chapter 6 focuses on a specific disease, leprosy, and examines the local meanings/representations of the disease and the sorts of actions authorized by cultural readings of it. Khmers had a rich constellation of definitions and medical techniques for leprosy, a disease that existed in the country well before the colonial era. Although leprosy was relatively infrequent, it was of disproportionate concern to French administrators. The chapter narrates the attempted transformation of an autochthonous leper village into a French *leproserie*, which in its ideal state was to be an efficient, rational totalizing institution. The failure of the institution, as well as the failed transformation of its occupants into model citizens, illustrates conflicts in representations of authority and the vast gulf between discursive colonial representations and the social reality they masked.

The concluding chapter, "Cultural Insolubilities," asks whether an effective Western medical system actually was constructed in Cambodia by the end of the colonial period. It calls into question how and whether scientific discourse is operationalized across discursive divides and different regimes of value. By the Vichy era, the colonial government had made little headway in "modernizing" Khmer understandings of medicine, although the French did create a rudimentary institutional infrastructure for Western medicine in the country. Ultimately, French technologies failed to fit in with Khmer philosophies of health and illness. The multiple narratives of medicine, derived in part from cultural insolubility, qualify the claims of universalism in Western medical sciences and of syncretism in one Southeast Asian society. Techniques to treat the body remained in discrete camps; basic metaphysical assumptions on the body, health, and disease remained distinct.

ONE

# Settings

Some time ago, inhabitants of a slum area in the capital city of a small kingdom informed the city's municipal doctor of a plague death in their quarter. When the doctor arrived at the home of the deceased, he found several pigs roaming near the wooden shack of the victim. The city's mayor, upon having this scene described to him, ordered the killing of all of the pigs in the city. This caused uproar from the general population, as pigs were a vital economic safety net for the poorer inhabitants. Despite a royal ordinance forbidding unauthorized travel during this time of epidemic, thousands took flight, many with their threatened livestock. Others gathered in a mob before the royal palace, pleading for assistance and redress from the king. The king came forward and assured the crowd that the killings had been in error and would cease immediately. This was the first encounter any of the participants had with this new disease, the bubonic plague.

There are two aspects of this story that make it a historical puzzle as much as a colorful vignette. First, the time and place: it happened not in a medieval Europe of disease etiologies based on miasmas and moral failings, but in colonial Southeast Asia in the early twentieth century, when scientific medicine was ascendant. Secondly, the actors' roles invert our standard narratives of science in the colony. In this instance, the doctor and the mayor were Frenchmen: Westerners, colonizers, modern. The king and the protestors were Khmer: indigenous, colonized, traditional, even "primitive." Yet, the Khmer king—we may say inadvertently—was closer to "modern" opinion in calling these pig killings an error and suggesting that they cease immediately.

As for the French mayor, he would write in his own defense:

> If [the pigs] had absorbed the feces or other contaminated materials fallen from this hut, they would be propagating agents of this evil, either in their

> waste, or through their consumption [by humans] . . . . Pigs being free to roam in this part of the town, [and] these pigs having surely mixed with the others . . . I adopted a radical measure; I ordained the slaughter of these animals and their incineration.[1]

Unfortunately for the mayor, his centuries-old reasoning was five years out of date.[2] Although as recently as 1900 prominent plague researchers would have agreed with the mayor, by 1905 scientific experts had established the link between the plague and rat fleas.[3] Unfortunately for the mayor, scientific medicine had also determined that pigs played no part in the plague's transmission. As the health director of Cambodia would sardonically observe some time later, an instance such as these "ridiculous" and "untimely" pig incinerations would not be repeated.[4]

This "ridiculous pig incineration" serves as a marker of several ruptures in the making of the modern conception of epidemic diseases. In the decades around the turn of the twentieth century, researchers and policy makers around the globe were locating, defining, and learning to manage diseases in a systematic fashion. This book is concerned with the transformation of disease definitions—medieval plague into the modern plague, consumption into tuberculosis, leprosy into Hansen's disease, and so forth—and the conflicts that arose as diseases were made "modern" in biomedicine. But medicine was not simply moving through the Western historical narrative; it was also being transplanted into non-Western settings, where other transformations took place. At the turn of the twentieth century, the French colonial government along with its scientific and medical communities was attempting to bring rational scientific understanding and practice to the entire domain of medicine in its colonies. This is thus also a story of how a culture vastly different from the defining culture in French colonial Cambodia attempted to negotiate the changing definitions of health and illness. It is a story of how concepts of health and disease mediate the relationship between biological, political, and cultural selves. With the yawning sociotheoretical gap between the practices of biomedical doctors and their local interpretation, transformations in the medical domain were not to be found in simple technological exchanges. Furthermore, during this key historical moment, the gulf in epistemological understanding between colonizer and colonized grew immensely with the development of microbiology and the revolution in germ theory in the West. But it was also at this moment that the French colonial government was attempting to universalize a specific understanding of medicine. French efforts to transform medicine in Cambodia would follow several trajectories during the colonial era; however, the results they obtained were never quite as intended.

## Colonization

Cambodia is a small country in mainland Southeast Asia, sandwiched between Thailand to the west and northwest and Vietnam to the east. Its northeastern border is shared with Laos. The ethnic majority of Cambodia, the Khmers,[5] reached their political apogee in the twelfth and thirteenth centuries, when the Khmer kingdom of Angkor encompassed portions of what are now Thailand, Laos, Vietnam, and Burma. Since the thirteenth century, the territorial reach of the Khmer kingdom has steadily decreased. Beginning with the writings of the colonial period and moving into the contemporary age, this decrease or decline has been conflated within Khmer history and historiography to speak not only of territorial reach but also of political effectiveness, cultural vibrancy, societal adaptiveness, and indeed racialized biological fitness (in the evolutionary sense). Although today we can perhaps eliminate the last item from this doomed cluster spiraling downwards into a sort of ethnic oblivion (or at best a Cham-like territorial dispossession),[6] other aspects of this notion of decline are deeply internalized in contemporary Khmer elite understandings of their own history as well as in Western historiography of the country.[7] It is difficult to assess how the Khmer elite perceived history (or the anachronistic and relativistic term of "historical evolution") at the point when the French began reshaping and reinterpreting the Khmer past in the late nineteenth century. Scholars today still operate within this framework of interpretation. While Khmer and some Southeast Asian regional studies may sometimes suffer from a somewhat provincialized interpretation of events, they do not lack for temporally expansive studies that by their very structure emphasize this decline and the "stagnant" nature of Khmer society.[8]

Although it is indisputable that Cambodia as a sovereign political entity was under serious threat when the French intervened in the second half of the nineteenth century, this situation had existed for centuries. The instability of the region was due to several factors, including shifting trade patterns that facilitated the rise of powerful kingdoms to the east and west, the relationship between the traditional Khmer elite classes and the general population (the so-called contest state),[9] and the absence of clear rules of kingly succession. Khmer kingship, in common with most of precolonial Southeast Asia, was not determined by primogeniture.[10] The king maintained a large number of wives and any of his sons from these various wives had a legitimate claim to the throne.[11] Possible legitimate heirs to the throne could number in the dozens, and death frequently took the king unexpectedly. Cambodia's geographical position between two rising political centers exacerbated the political instability.[12] Ultimately, this combination of factors led to centuries of civil war over kingly succession that wreaked havoc on the region's governance. European attempts to take advantage of the perennial instability began a century after the fall of Angkor, when Spanish and Portuguese adventurers partook in several misadventures interfering with royal politics.[13] In the eighteenth and nineteenth centuries, the factions vying for Khmer kingship alternately looked to their increasingly powerful neighbors to the east (Vietnam[14]) and the west (Siam, modern day Thailand) to support their claims to the throne.

A history of Cambodian kings immediately before French intervention reveals this pattern. The Khmer kings of the turn of the nineteenth century, King Ang Eng and his successor Ang Chan, were both crowned in Bangkok and claimed the throne of Cambodia with the aid of a Siamese army. Throughout his reign Chan sent tribute to Bangkok, but in 1811 he turned to Vietnam for assistance against a brother vying for the throne. This led to a destructive battle between Siam and Vietnam fought on Cambodian soil. Ultimately the Vietnamese prevailed, forcing Chan's half brothers Ang Em and Ang Duong to flee to Bangkok. Chan, with Vietnamese support, was to reign until his death in 1835. At his death he left no heirs, and Vietnam attempted to annex Cambodia as a sort of colony, naming Chan's daughter Mei queen. Under the reformist policies of Ming Mang, the Vietnamese king, the Vietnamese government imposed drastic changes on Khmer society in an effort to make it conform to a more Sinicized Vietnamese cultural and political unit. Between 1836 and 1842 the country witnessed a growing rebellion by Khmer peasantry against Vietnamese administration. By 1840, Ming Mang was facing a countrywide revolt.[15] Whatever the main engine for the revolts, French scholars would note that the Vietnamese atrocities under Ming Mang's reforms were still vivid in the minds of the Khmer peasantry at the end of the nineteenth century. As French explorer Louis de Carné observed, from this period forward "the dislike, which has always divided the two races [Khmer and Annamite], changed, on the one side, into an inextinguishable hatred, on the other to a profound contempt."[16] Memory of this epoch would prove to be long, despite—or perhaps because of—little formal written history. This enmity between the Vietnamese and the Khmers would be exploited and exacerbated by French rule in Cambodia, and would impact the development of Western medicine and racial identity in the colony.

In 1841, at the height of the Khmer revolt, Siamese General Bodin and his armies accompanied Prince Ang Duong into Cambodia to wrest the throne from the Vietnamese-supported Mei. After years of skirmishes, Siam and Vietnam signed an 1846 treaty crowning the Siam-supported Ang Duong as the king of Cambodia in 1847.[17] The Vietnamese interregnum damaged relations between ethnic Khmers and Vietnamese; this discord would initially be exploited by the French but would later create serious problems for the French medical system in Cambodia.

With Duong's death in 1860, the kingdom predictably erupted into another civil war. Before his demise, Duong had named his oldest son Norodom to *obyureach* (presumed heir). Despite this, Norodom's two half-brothers immediately challenged his claim to the throne. One half-brother, Si Votha, turned to the provincial heads and Vietnam for support. Once again, the presumptive king—this time, Norodom—was forced to flee to Bangkok. Like his father, he soon returned with the aid of a Siamese army. Siamese assistance to the Khmer crown deeply worried the French, who suspected British intrigues in Siam; this was to be a major French pretext for involvement in the country. The other major claimant to the throne, Norodom's half-brother Sisowath, who was labeled ignobly but perhaps justly as a French puppet, came to play a very important role later in the French colonial period.

Much quality scholarship already exists on the motivations for French involvement in Cambodia and insular Southeast Asia more broadly. It is generally agreed that French adventurers, speculators, soldiers, and missionaries in the region took the initiative on behalf of their mother country in imbricating France in local politics. The close, if unofficial, rapport between the navy and missionaries was key to military and ultimately political intervention in the region.[18] In the first half of the nineteenth century, French Catholic missionaries and their converts were on occasion intensely persecuted by the Vietnamese government. French missionaries were eventually successful in securing promises of aid against this persecution from the French naval commanders stationed in the South China Sea. Religious sympathies would mesh with imperialism, nationalist sentiment, and economic considerations to impel intervention in the Mekong Delta.[19] Napoleon III gave the French navy permission to attack Vietnam in the area of Tourane (Da Nang) in 1858, ostensibly to aid persecuted Vietnamese Catholics. Although assisted by Spanish Catholic sympathizers, the French navy was unsuccessful in holding Tourane, and moved further south to Şaigon, an area it secured in a decisive 1861 battle. After his forces were defeated, the Vietnamese king at Hue signed eastern Cochinchina over to French control in the Treaty of 1862.

Shortly before the outbreak of hostilities in Vietnam, Norodom had approached the French government with interest in establishing some sort of tributary alliance. Most likely he viewed France in a similar way to Siam or Vietnam, as one strong power to play against the others. In this, he would be gravely mistaken. Shortly after securing Saigon, the French representative at the court of Oudong in Cambodia concluded (or, according to some histories, "coerced") a protectorate treaty in August 1863.[20] This initial treaty gave France timber and mineral concessions in exchange for its ill-defined protection against other interests. From the signing of the 1863 treaty, French administrators slowly encroached upon Norodom's powers, expanding unilaterally upon the rights originally vested to them. In 1884 the French regional governor of Cochinchina, Charles Thomson, forced the king to sign a treaty greatly expanding the power of the French protectorate over the country. The 1884 treaty turned Cambodia for all practical purposes into a full-fledged colony. It is only a slight exaggeration to say that the king was forced into this action while looking down the barrel of a gun, as Thomson took the precaution of bringing in three gunboats from Saigon and leveling them on the palace before negotiations began.[21]

Much to the surprise of the French administration, a countrywide revolt erupted immediately after this treaty was signed. The protectorate was forced to call upon the assistance of King Norodom—who was suspected of instigating the uprising—to quell the revolts. Using both threats of deposition and promises of reward, the French representative persuaded the king to urge the populace to cease hostilities. Ultimately, the king did cooperate in asking the population to lay down its arms. In return for Norodom's aid, French administrators temporarily rescinded some of their more egregious political and economic usurpations. Although scholars fault the French for being very slow in investing in the "improvement" of the country through schools, infrastructure, and economic development, this slowness of the French was due in part to their hesitation to push the uncooperative Norodom too far and risk spurring another insurrection. The colonial administrators preferred instead to wait out what was expected to be the imminent demise of the sickly king. Although constantly ill, Norodom surprisingly lingered on another twenty years into the early twentieth century. With his death in 1904 and the ascension to the throne of his half-brother Sisowath—hand-selected by the French—colonial involvement in Khmer government and society rapidly increased.[22] It was during the Sisowath years (1904–27) and those of his son Monivong (1927–41) that the French economic and political reach into the Cambodian countryside rapidly grew. The bulk of this study concerns this period of rapid economic development.

## Colonial Administration

By 1884, France had claimed the lower Mekong Delta. By 1893 it had also obtained control of Laos, central Vietnam, and northern Vietnam (Annam and Tonkin), and proclaimed the five "states" of Annam, Cambodia, Cochinchina, Laos, and Tonkin the united political unit of French Indochina, administered by the Gouvernement Général de l'Indochine (GGI) under the authority of the French minister of colonies.[23] The governor general directly administered Cochinchina and supervised the French heads appointed to the other four states, the *résidents supérieurs*. The *résident supérieur du Cambodge* (RSC) governed from the new capital city of Phnom Penh. Cambodia was further divided into several districts, each administered by a French representative, the district *résident*.

Cambodia as a protectorate was under a system of indirect rule, which implied that indigenous systems of administration were to be minimally disturbed. In this, it was similar to three of the other four states of Indochina.[24] France was to represent the country in its international endeavors and advise it on domestic policies, but domestic governance was to remain a native matter. In reality, the colonial government quickly encroached on all fronts in Cambodia: domestic, international, economic, and administrative. Ironically, although the actual political power of the royal court and the indigenous administration decreased over the colonial period, the significance of the king grew in the minds of his countrymen during the same period.[25] The Khmer Council of Ministers in conjunction with the king continued to serve as the ruling government. Importantly, the French résident supérieur du Cambodge now sat as the president of the Council of Ministers. Through the first decade of the twentieth century, this council had a significant role in setting internal policy and advising the colonial government. By World War I, however, its role and that of the king devolved to little more than rubber-stamping French colonial decisions.

The official period of French colonial rule was 1863 (greatly expanded in 1884) to 1954.[26] However, this study focuses on the French medical services in the country and their impact on the indigenous population during the years 1907 to 1940. The colonial government, with a few exceptions, made no concerted efforts to provide Western medicine to the indigenous population until 1905, when it decreed a native medical service. The Cambodian service did not begin operations until 1907. With the advent of Vichy in France in 1940, the colonial medical services were severely disrupted, many aspects suspended or eliminated. The medical services never truly resumed, although France attempted to reclaim its role as a colonial power at the end of World War II. Its attempts were not successful, as the First Indochina War (between France and Vietnam), postwar reconstruction at home, and the rising tide of anticolonialism quickly absorbed the attention of the metropole. In 1954, during the Geneva Convention, France relinquished control of Cambodia to young King Sihanouk, son of Monivong.

## The Khmer Social Milieu

In the late nineteenth century, the majority of Khmers were subsistence farmers. In times of peace, which had been rare, most Khmers seemed to have lived a relatively uneventful life marked by the changing wet and dry seasons and the many periodic religious ceremonies.[27] The country had few large towns, and thus few urban dwellers. Much of the urban population was non-Khmer. In the beginning of the colonial period, the connection between the capital and rural towns was sporadic.[28] Travel routes were rudimentary; flooding and erosion during the monsoon season (May–November) made many land routes impassable. Travelers depended upon one of the many water routes that crossed the country in the wet season, and those living away from a major body of water were greatly isolated. Even during the dry season, travel was difficult. Perhaps for this reason, Khmers were resistant to relocation and unnecessary travel. Although the study will examine this further, it seems that the lateness of the expanding road and railway networks in the colonial period meant that those routes had only middling impact on theretofore village life. Considering that the majority of Khmers existed with little contact with the outside world, it is unsurprising that we have few firsthand accounts of their lives. This is compounded by the fact that most rural Khmer were illiterate, and secular writing was sparse.[29] With the partial exception of the royal chronicles and transcribed oral poetry,[30] most written materials from the late nineteenth century are religious in nature. Folktales, inscriptions, *krang*, and palm-leaf manuscripts comprise alternative historical sources. From these materials, as well as from European accounts, studies of neighboring societies, and materials from the postcolonial period, we can glean some generalizations.

Rural Khmers were surrounded by an often hostile wilderness. Urban and rural Khmers made a strong distinction between wild and cultivated, or *prey* (wilderness) and *srok* (town, province, country, etc.).[31] The wilderness not only was a place of uncivilized bandits, but also contained tigers, crocodiles, elephants, rhinos, and other dangerous animals.[32] Further, it hosted countless amorphous fey spirits and ghosts that were unpredictable and often intent on causing harm to humans. As one anthropologist writing in the 1940s described a Cambodian hill tribe: "Every single fiber of their being rooted in a hostile milieu of forests and mountains, these autochthones live in anxiety and fear of the forces of nature, which they personify in immaterial form."[33] Theravada Buddhism was the predominant organized religion; most adolescent boys spent a token amount of time within the Buddhist sangha (monkhood) to build merit.[34] The indigenous population also followed a variety of animistic beliefs. Animism, which Paul Mus termed the pan–Southeast Asian primordial chthonic religion,[35] attributes deities and spirits to both animate and inanimate objects. In the Cambodian village, the most important of the animistic spirits was the *neak-ta*—literally "the grandfather," though it could be either male or female (sometimes called *neak-yey* or "grandmother"). The *neak-ta* was usually a protector spirit. Some villages had a communal *neak-ta*, others had family or site *neak-ta*, and many villages had both. If angered, the *neak-ta* could cause his village misfortune. He or she was frequently called upon to appease other types of spirits that had brought sickness or bad luck.[36] Magic, sorcery, and what we would call superstitions were strongly integrated into daily life through ceremonies, rituals, taboos, and myriad customs.

## Indigenous Medicine

The magical and supernatural was also strongly integrated into indigenous medicines. Again I emphasize that this study is *not* constructed as a comparison of two epistemologies of medicine. A history of one system does not naturally include a history of the other, as an understanding of Khmer medicine leads one to realize that often the French doctors were not offering the indigenous population what it would call medical care. In chapter 3 I will define specific aspects of French medicine in relation to the variety of indigenous medical practices it confronted. Here, it is useful to highlight key features of Khmer medicine that provide a background for the reasons why the indigenous population reacted as they did to French medicine. These descriptions are informed by the anthropological perspective that medical "meaning and knowledge are always in reference to a world constituted in human experience, formulated and apprehended through symbolic forms and distinctive interpretive practices."[37]

Several different ethnic groups existed in Cambodia, with wide-ranging medical beliefs. For example, the sizeable Cham population had certain practices based on Islamic principles. The Sino-Khmer and Annamite groups focused more strongly on medical practices derived from Chinese medical traditions. These three groups practiced differing forms of smallpox variolation (nasal swab, scarification). In contrast, Khmers often did not practice any form of smallpox variolation.[38] However, the many ethnic minorities in the country seemed to hold in common with the majority Khmers an expansive, general system of understanding of the human body and the etiology of disease. This system, or these systems, derive from animistic, Buddhist, and Hindu philosophies. The most important components are essences and spirits. All humans have nineteen *praluung* (vital spirits or consciousnesses) and four *cato phut* (major essences). *Cato phut* is a corruption of two Sanskrit words meaning "four elements"; the theory is likely of south Indian (Ayurvedic) origin. The *praluung* and similar iterations appear to be specific to mainland Theravada Buddhist cultures.[39] These elements of the human spirit, particularly the *praluung*, are inclined to wander off or be called away. If one or more of them is absent for an extended period of time, the person will fall sick and a ritual ceremony is required to call the vital spirit back to its host. The absence of a *cato phut* could be more serious. Further, if the *cato phut* returns to find its host dead, it will attempt reincarnation as quickly as possible and is prone to become an errant spirit attached to rocks, trees, or other natural elements.[40]

Theravada Buddhist medicine, Ayurvedic medicine, and animistic medicine each have an independent and complex set of representations of the body and its functions.[41] However, Khmer traditional medical practitioners and their patients did not carefully distinguish or categorize the type of medicine they practiced.[42] A notable exception was Buddhist medicine, some of which was derived within Theravada Buddhist philosophies that were linked to the institutional framework of the sangha. But even medicine practiced by Buddhist monks often incorporated animistic beliefs. A literate medical tradition existed, most notably in the form of palm-leaf manuscripts, which consisted of knowledge bound in sets of treated palm leaves. A review of these manuscripts reminds one that the classification of disease can be an arbitrary affair. One set of palm-leaf manuscripts may give the medicinal treatments for seizures, unusual tissue growth on the nose, giving birth, and blindness. Another set could provide invisibility spells, techniques for spitting on sores, and cures for venom bites. A cure for leprosy could follow a recipe for black hair dye.[43] The Buddhist sangha employed palm-leaf manuscripts, but some learned folk practitioners also used them to record their medical knowledge. The various representations of the body that could be extracted from different philosophies, even if contradictory, were not mutually exclusive. This was perhaps a reflection of the syncretism that is said to bind the Southeast Asian region. However, the multiple viewpoints may also be due to conflicting Western interpretations.

Ultimately, sickness was caused by supernatural as often as natural entities. Supernatural causes of disease could range from malevolent spirits[44] to witchcraft.[45] Etiologies that were more difficult to classify included the various taboos against certain foods or behaviors in specific settings or times, as well as missing *praluung*.[46] *Kchaal* (breath or wind) played a vital part in human well-being.[47] Many illnesses could be caused by disruption of the balance or flow of *kchaal* in the body. Some *kru* believed that the network of *sasae* (threads or vessels) transported the *kchaal* throughout the human corpus.[48] Weakness in the *sasae* could further cause one to fall sick. Massage was often used to reinvigorate blocked or fatigued *sasae*. Illness could also simply be caused by an accident or other natural means.

Just as disease etiologies incorporated both the natural and the supernatural, treatment did as well. Treatments for serious diseases usually required some form of ritual. For more minor ailments, most people self-doctored as a first option. As in all cultures, the sick employed folk remedies for these minor ailments before consulting experts. One of the most common home remedies for sickness was coinage, where a coin dipped in lubricant was scraped repeatedly in stripes across the body to correct the internal imbalance in *kchaal*.[49] Most Khmers wore (and many still wear) a ritually blessed protective string around their waists to ward off diseases.[50] Ritual blowing by a monk or sorcerer on the object was central in blessing such wards.[51] Whether the healer was a spirit medium or an empiricist, he or she almost always had to perform some sort of ritual action with the treatment. For example, an 1890 account of indigenous Khmer smallpox treatment describes the healer covering up the largest sores with a pulpy mixture consisting of datura root, mushroom, and young palm stems. While applying this to the pox on the body, the healer chanted, "I do not fear you, but if you were on the shoulder, I would fear you."[52] This chant was intended to lure the smallpox, a sentient being, off of the trunk of the body to the extremities. As with many diseases, the specific diagnoses and treatment of smallpox depended on the day of its onset.[53] Also, as with many other diseases, the patient's astrological chart could be significant in predicting the severity and appropriate treatment.[54]

The varieties of medical practitioners, like those of medical systems, resist a penchant for exact classification.[55] Power, merit, and right action were also intricately linked with the ability to heal. Not simply possessed of secular qualifications, a healer was the triune of the moral/meritorious/powerful person. The source of his power paralleled the source of disease. Order in the cosmos, order in society, and order in the individual were inextricably connected. Thus, healers and holy men were often linked. The *achaar* (roughly, a lay officiant) of a *wat* (Buddhist temple) or a particularly skilled Buddhist monk often developed a reputation as a healer. Further, monks frequently participated in ceremonies to ritually ward objects or locations against bad luck and disease. Healers could have very narrow abilities—for example, they could specialize only in "spitting" to cure red eyes or in providing a specific herbal remedy for leprosy—or they could be capable of treating the entire spectrum of human discomfort. They employed a vast array of natural products, ranging from plant bark and animal organs to charcoal, in their treatments.[56] Some healers specialized in pharmaceutical products to alleviate symptoms, with the necessary supplication to offending spirits. Magical dreams often revealed the healing properties of natural products to those meant to be healers. Along with personal revelation, medical knowledge could be a familial affair, or a gnostic process of instruction from teacher to student. Other healers did not deal at all with medications, serving rather as spirit mediums (*ruup*) or sorcerers to negotiate directly with the spirit causing the disease. Other individuals, not exactly healers, would claim to know or possess objects that healed or made people impervious to disease. The most common such object appears to have been lustral water.[57] Tattooing was also a common form of "immunization" against injury and disease, although it was likely an exclusively male practice.[58]

The current Khmer translation of the English term "doctor"[59] is *kru peet*. *Kru*, derived from the Sanskrit term *guru*, denotes teacher or learned person. The term *peet* is translated as hospital or clinic. Today, to differentiate a traditional doctor from a Western-trained doctor, the traditional (and popular) healer is referred to as *kru khmer* or simply *kru*, and a Western-trained doctor as *kru peet*. It is unclear when the combination of the two terms "*kru*" and "*peet*" was first used to denote a Western doctor. However, a 1965 study made a vague claim that a Khmer doctor—in the original French, *médecin khmer*—who practiced traditional medicine was a *kru peet*, while "popular"-medicine practitioners were in a different (unnamed) class, presumably just *kru*.[60] Practitioners of traditional medicine were *kru peet* because they depended on a written, learned tradition based on palm-leaf manuscripts. Practitioners of popular medicine were more informal, self-trained empiricists. A later researcher argued that little distinction could be made between practitioners of traditional and popular medicine: both kinds of practitioners were simply called *kru* or *kru khmer*, and neither depended heavily on palm-leaf manuscripts.[61] An indigenous colonial doctor would claim in the late 1920s that only aged Khmer healers consulted the sacred medical book written by the Buddha.[62] Although in recent years concerted efforts have been made to record and preserve palm-leaf manuscripts,[63] they have not yet been systematically studied.[64] Regardless of whether *kru* and *peet* were commonly used together to denote a specialized healer in the Khmer context before the colonial era, the evolution of this compound term exclusively to mean a Western-trained doctor could indicate something of the institutional associations of Western medicine. Since the definition of *peet* is linked to a type of institution, the use of *kru peet* in preference to other compound terms such as *kru mul chomngu* (*kru* who looks at sickness) could reflect differing associations of healers. In other words, the *kru peet* practices in an institution. However, this is difficult to determine with any certainty.[65] French doctors are not commonly referred to as *kru peet* in documents disseminated in Khmer until late in the colonial period.[66] This slippage in language is just one of several French-Khmer (to English) translations relating to medicine that muddles apprehension of the transformations in this period.

The range of healers in Khmer society varied, and this was complicated by the differences between types of healers among the different ethnic groups. Some healers were purely magical, others were empirical, but most seemed to be a mixture of both. The *kru khmer* were often further distinguished by their specialties (the spitting *kru*, the bone-setting *kru*, the love-potion *kru*, etc.). When the French arrived on the scene in the nineteenth century, the country had no *peet*. In other words, no establishments existed that were dedicated predominantly to treating the sick. The sick were treated in their homes, in the village centers, at the homes of healers, or at the *wats*. Sick patients could also be transported to sites of reputed magical healing power, such as a tree or shrine.[67] However, *peet* had existed before in the country. Much has been written of Jayavarman VII's progressive state sponsorship of hospitals during the height of Angkor's power in the early thirteenth century.[68] In operation, these hospitals were much more than sites for medical care. At a given site, along with the two allotted doctors, a supporting staff of orderlies, assistants, guards, cooks, and supplicants could number well over one hundred individuals.[69] It could be argued that these hospitals, placed throughout the kingdom, had some similarities with European almshouses of the same period. But these *peet* had long since disappeared before the time of our study, and in any case they were radically different from the evolving notion of the hospital that the European brought at the turn of the twentieth century. Given the scant Khmer textual or archaeological evidence on the subject in the intervening centuries, and the absence of hospitals at the arrival of the French, it is difficult to assess what the term *peet* meant to a Khmer person during the period of our study. Certainly, the Khmer peasant did not view care of the sick as a governmental concern. And, even with the earlier definition of the hospital, we can assume that sickness and its treatment were seen as a socially embedded affair. The entire community or the village *wat* could be intimately involved in healing rituals for a sick individual. This would be in stark contrast to the French view of the hospital as a site of separation of sick individuals from society, where they were placed under the authoritative management of a few.

In sum, Khmer medicine is not a neatly categorizable affair. Its philosophical underpinnings ranged widely. To a greater extent than within late-nineteenth-century Western philosophy, the epistemology of Khmer medical knowledge could not be delineated from other natural philosophical systems. The majority of healers did not have a textual basis for their craft. A written tradition did exist, but it was based in large part on the Buddhist sangha. If a standard written tradition of medicine is necessary to codify medical practice and theory,[70] it would seem that the best analysis of Khmer medicine in the French colonial period ought to remain in the realm of generalization simply because of its fluid nature and wide range of practices. Any assessment of doctors would also have to consider that the range of healers and the authority granted to them varied widely. As Prince Ritharasi Norodom would observe in 1929, to be a doctor in Khmer society, one merely had to claim to be a doctor.[71] It must be added, however, that the respect accorded to the doctor depended directly on his abilities either to convince or to heal. Almost no authority was given solely to the title or role of doctor; authority existed through the reputation garnered by the practitioner within the population. This would be in stark contrast to the institutional authority presumed by the incoming French doctors.

## The French Medical Corps

The outlines of French medicine as a subset of Western medicine at this historical moment will necessarily be grossly oversimplified in the few paragraphs that follow. Unlike indigenous medicine in Cambodia, it had a much stronger institutional and state base. We can trace the modern shape of state regulation of French medicine to reforms after the fall of the *ancien régime*.

As medical historian Jacques Léonard has observed, the nineteenth century began with Napoleon's France ushering in a scientific and biopolitical revolution in medical practice, and ended with a realist France that enthroned Pasteurian experimental medicine.[72] French medicine as theory, as practice, and as philosophy rapidly evolved during this period. The wide-ranging transformations are beyond the scope of this book; others have written on various subtleties of this nineteenth-century medical revolution.[73] Changes that were established earlier in the nineteenth century—the strong institutional base, state involvement, and professional status of French medicine—would already be in striking contrast to medicine in Cambodia. Of greatest concern to us are the areas being contested during the period of our study. Although many aspects of medicine were in flux during the early twentieth century, we can identify two major sea changes in French medicine that would strongly influence its development in the colonies: the rise of germ theory and laboratory science, and the increasing application of population thinking to health and medicine.

The French medical services in Cambodia began in a haphazard fashion. The first French doctor who served both a French and Khmer political role appears to be a Dr. Hennecarte who provided medical advice to King Norodom at the royal court of Oudong in the early 1860s. The French navy had a dedicated corps of military doctors whose functions could include monitoring the health of French troops, carrying out exploratory expeditions, or performing anthropological work. Indigenous health, however, did not fall within the scope of its duties.[74] The large number of casualties resulting from the 1884 uprising impelled an improvement to the makeshift Phnom Penh ambulance. Under the direction of naval doctor E. Maurel, the Phnom Penh military hospital was founded.[75] This would later become the premiere hospital in the country for both natives and foreigners. At its inception it was designed only for European soldiers, with a small section allotted for Vietnamese soldiers brought in to fight against the Khmers. By 1886 the medical service consisted of one semipermanent establishment (the Phnom Penh ambulance), and doctors attached to four military regiments around the country.[76] With the partial exception of the Catholic missionary crèche built in 1881, Western medicine was at this point almost exclusively for Europeans.[77] In 1890 the French metropolitan government created a Colonial Health Service (Service de Santé des Colonies) separate from the military (marine) medical service. In conjunction with this, the government also created a Colonial Health Corps (Corps de Santé Coloniale). A new training institute dedicated to this service, the Ecole de Santé Navale, had opened in

Bordeaux, France, in 1887.[78] The Colonial Health Corps existed in a murky form until it was definitively reorganized as the Health Corps of Colonial Troops (Corps de Santé des Troupes Coloniales) in 1903.[79]

Despite the reorganization of colonial health services, the high mortality and morbidity rates of soldiers and settlers remained a serious detriment to effective colonization.[80] Government efforts to combat diseases eventually had to extend to the indigenous populations, who were seen as disease reservoirs. Further, the humanitarian philosophies buttressing liberal ideals of colonization demanded that all of the benefits of "civilization," including better medical care, be made available to indigenous populations. Each colonial power in Southeast Asia eventually came into line with this thinking and extended Western medical services to conquered populations.[81]

Mobile vaccination runs represent the first colonial efforts at targeting the health of the indigenous population. Mobile military doctors began performing vaccinations in the hinterlands of Indochina in the 1880s. This was quite a puzzling behavior to the indigenous population; the French vaccinator would be the first and only Western contact for many of them in this early period. In 1892, a Khmer interpreter working in the remote region of Kratie (in northeastern Cambodia) asked the French explorer and Pasteurian Alexandre Yersin if he planned to vaccinate on his next exploratory journey. When Yersin answered in the affirmative, the interpreter reportedly said to a fellow aide, "These French are really shocking, and all the same!"[82]

Effective vaccination programs ultimately required reliable vaccines. To this end, in 1890 the new Colonial Health Services, in conjunction with the newly created Pasteur Institute and the also relatively new Indochina government, sent the soldier, medical doctor, and Pasteurian Albert Calmette to create a vaccine institute in Saigon. Calmette would be the first of many basic researchers in Indochina linked to the Pasteur Institute. These Pasteurians and their research programs would have a substantial if indirect impact on French medicine in Cambodia.

Along with vaccination, organized health programs for the native populations were soon to follow for all French colonies. The first French colonial health program for colonized populations, the Assistance Médicale Indigène (Native Medical Aid), was formed in the French African colony of Madagascar in 1903.[83] The following year, the metropole issued a general health decree ordering the creation of medical services for indigenous populations in all colonial possessions.[84] For Indochina, a decree of June 30, 1905, created the foundations of the Assistance Médicale Indigène de l'Indochine.[85] As originally decreed, the director of health services of occupational forces was named as the inspector general (*médecin inspecteur*)[86] of the Assistance

Médicale (hereafter AM). Already charged with the doctors of the colonial troops, his duties were extended in anticipation of the new, incoming corps of civilian doctors. This co-appointment of the military health director as the civilian health director was considered an exceptional circumstance. As the first inspector general of Indochina explained of his dual role:

> This measure was indispensable for permitting the success of the oeuvre undertaken and to provide the same direction to the diverse factors that collaborate with the Assistance: military doctors, civil doctors, and indigenous doctors . . . . [T]his [method of] organization has greatly facilitated the performance of the service; it allowed us to avoid learning by trial and error [*tâtonnements*] and the [other] difficulties of a debut and it allowed a coordination that could not have been obtained under other conditions.[87]

The inspector general oversaw the local health directors (directeurs locale de la santé or DLS), one for each of the five administrative sections of French Indochina.[88] The powers of the inspector general were not particularly well defined, particularly in relationship to the colonial government. Although he was to set the general direction of the health services and advise on promotions and placements, he had to bow to the authority of the governor general, who had ultimate authority to hire or fire the local health directors or their staff. The governor general not only had the final say on the health budget and issued administrative decrees for the AM, but also, in conjunction with each résident supérieur, was the immediate superior of each state's local health director. Thus, the powers of the inspector general were to fluctuate with the goodwill of the governor general. Military, scientific, and civilian priorities were often in conflict. The director of the Indochina Pasteur Institutes would have similar conflicts with the colonial government. He, too, would negotiate his role under the triple authority of the Parisian Pasteur Institute, the inspector general of the Health Services, and the governor general of Indochina.

The Cambodian sector of the Indochina AM was not launched until 1907. At that time, Cambodia was divided into several medical districts (*conscriptions*), which numbered at any time between nine and sixteen. Each district had a medical chief, who was based at the same locale as the French regional administrative director, the résident. Initially, the local health director[89] of Cambodia was also the chief medical resident of the Phnom Penh Mixed Hospital. Eventually, those duties were officially separated. Within a few years of the inauguration of the AM in Cambodia, district medical chiefs were required to submit monthly and annual reports to the health director.[90]

The health director then submitted an annual report for Cambodia to the inspector general, who provided a final report for all of Indochina to the minister of colonies.

As the AM matured, the organization expanded with mobile medical teams, hygiene groups, public health divisions, and so forth.[91] Again, although the AM was not a branch of the military, almost all of its doctors were initially military doctors. These positions were not terribly advantageous for the military doctors. At the AM's inception they were suddenly charged with the services of the AM along with their preexisting military duties. Although these men were technically on special assignment to the AM (*hors cadre*) and were placed on leave from their military service, in truth they continued to serve in both civilian and military capacities simultaneously. On average for this expanded duty, the government provided each of these men a supplement of 2,000 francs to an annual military income of 11,000 francs. This raised their salary to 13,000 francs per annum, making it on par with that of the average civilian doctor entering the AM.[92]

Military medicine, the AM, and the Pasteur Institute were the three main branches of French medicine operating in Cambodia. While some "branches" of Western medicine were not directly active there, the ties within the networks of medicine often meant that the influence and action seen in Cambodia could be far removed from the impulse that had created it. Thus, the next chapter contextualizes these different branches of the Indochina medical apparatus in relation to the indigenous populations, as well as their internal relations with one another, before focusing specifically on Cambodia.

TWO

# Collusions and Conflict

This chapter provides a structural overview of the French health services in Indochina, and of their position in colonial and international medical research. It characterizes the role of the men of the health service—and initially they were all men—within these fin de siècle scientific exchanges. The scientific enterprise within the colonial system has often been portrayed as a linear center-periphery (or "center–alternate center") model of knowledge production.[1] The colony is often seen as the site of data collection, while the center/metropole is the site of synthesis. When viewed from the colony, however, scientific systems appear as intricate networks and nodes of knowledge production. Something close to the final product is often determined within specific regions, with the metropole serving not as the synthesizer but rather as the communicator of this knowledge. Thus I would propose that in many scientific enterprises, the center does not produce the science; rather, it *sanctifies* the final product. Global dissemination of knowledge may be controlled at specific metropolitan hubs, but these hubs can be incidental to much of the work of research. Within the global context, the metropole is more accurately defined as a center of communication than a center of all major scientific production.

When we examine the three main branches of French medicine in colonial Indochina, we see that these actors did not perceive themselves as peripheral in research. On the contrary, many were on the cutting edge of scientific discovery. However, the story for practicing doctors was different. Many clinical practitioners believed themselves disadvantaged as career doctors; it was bad to be a doctor in Cambodia, better to be a doctor in Vietnam, and best to be a doctor in France. These clinicians working in the field, unlike the researchers, were often ill-equipped and less informed than their metropolitan peers. Being a member of the colonial administrative network was quite different from being a member of the scientific research network, although many men were both. How these branches interacted in relation to vaccine implementation programs reveals the complicated dynamics between scientific and administrative contingents within colonial Indochina and within global research networks. While the goals between these sectors often conflicted, disputes were largely self-mediated within the medical ranks. Further, the response to attacks from outside the scientific realm expose a remarkable political cohesion that existed among these men despite significant social and ideological differences. Political solidarity trumped scientific, social, and civic differences.

## Spheres of Authority

Although Cambodia was technically a protectorate nation, and thus in theory partially self-governing, by the second decade of the twentieth century the French controlled nearly all aspects of government. Unlike British imperialists, French imperial agents were predominantly soldiers, technicians, and teachers rather than merchants and colonists.[2] French historian William Cohen noted that much of the turn-of-the-century reform in French colonial administration, including the institution of the École Coloniale for administrative training, was an attempt to curb the rule of exceedingly brutal military administrators.[3] Again unlike the British colonial medical services, which in Africa were tightly tied to the powerful metropolitan British Medical Association,[4] the doctors of the French colonial medical service had no common professional association or union to represent or protect their interests. Their only comparable group membership was their tie to the military and, by extension, the colonial apparatus. Later in the colonial period, this connection would weaken. Further, members of the AM and the Pasteur Institute Indochina often shared a military background, especially before World War I. Thus the interests of the AM, the Pasteur Institute, and the colonial apparatus were bound together. Although, as we will see in this chapter, these groups did not always neatly agree in their goals, they often supported each other against outside attacks. Does this mean essentially that professional autonomy failed or did not exist in the colonial context, and that the medical prerogative was subjugated to colonial needs? Or did it give the hygienist, the scientific researcher, and the medical doctor greater freedom to extend their authority? Did the colonial apparatus provide a freer field of experimentation for the researcher while co-opting the physician as a tool of surveillance and control? Many of these readings overlap, and each is acceptable in specific circumstances.

However, any generalizations about these men require many qualifications. For example, the Pasteurian Alexandre Yersin would show remarkable callousness in complaining thus of his research assistant in India: "The Pondichery doctor, Bonneau, bothers me a bit. He is stupid and follows me like a shadow. He does not understand that I am performing experiments and shows too extreme a regret for each case that dies. This annoys me."[5] In stark contrast, in 1913 all but two French doctors in Cambodia refused to use the experimental drug Mycosiline on the indigenous population. One of the doctors wrote to his superiors explaining his refusal as a deliberate act of conscience.[6] For a scientist such as Albert Calmette, who was committed to the Pasteurian research program, the colony provided many more opportunities than metropolitan research to expand fields such as tropical medicine and antivenom experimentation.[7] For a researcher such as Yersin, the experience meant something entirely different; the colony fulfilled a romantic fantasy of travel while allowing him great freedom to avoid surveillance and to conduct work as he chose. For some practicing physicians posted out in the provinces, the colony was often only a job—one that led to a better posting elsewhere. For health officials in Cambodia, as we will see in later chapters, attitudes ranged widely. Ultimately, the doctors and medical scientists were individuals operating under a system that both constrained and enabled them in ways distinct from those of the metropole.

The medical researchers' main enemy, the epidemic disease microbe, had total disregard for geopolitical borders. The prestige that came with its discovery and control was also international. However, while scientists conducted research across national or imperial boundaries, their motivations were still tied to discrete affiliations. The community of science was not always communitarian. In this chapter we will discuss international research on two "colonial" diseases, the plague and cholera. While British metropolitan authorities were open in inviting all leading international biologists to conduct research in Britain's colonies, scientists and administrators could be quite territorial within the colonies themselves. British doctors in Hong Kong and India could be hostile and jealous towards foreign researchers. While scientists from Germany, France, England, Russia, Italy, and Japan were fervently conducting their research in the British colonies of Hong Kong and India, they were for the most part working independently of each other.[8] The international slant of research in the colonies intensified, at least for many French scientists, the tie between national pride and scientific research. Both successes and failures by a French researcher reflected on France; concealment of errors was in the interest of national pride. It was also in the interest of imperial pride, for the ability to prevent these diseases in the colonies was a reflection of competence as an imperial power through the ability to fulfill the promise of improving subjects' lives.

We can speak about the intentions of the "system administrators" in the metropole, but many historians have argued that policies in the colonies were often driven locally rather than from France.[9] In Indochina, we can identify the inspector general or the Pasteur Institute director (Yersin, perhaps, or Bernard) as the local policy makers, yet in truth they often had limited power or none, their prerogatives guided by the GGI or the Parisian Pasteur Institute. Often the governor general did not involve himself in the affairs of medicine; at other times the health service was pushed by trends in metropolitan science and society. Equally important, the health services may have appeared on paper as a product of these various factions, but it was also strongly guided or constrained by local conditions. These conditions were not only social or economic but also cultural, religious, and geographical. When talking about experimentation upon colonial "guinea pigs," we soon realize that these guinea pigs were not voiceless animals, but people making choices that the medical services and the colonial government were forced to address.

This chapter begins with the history of plague research as it developed in Indochina, moves to a global arena, and then returns to Vietnam and Cambodia. It then details the story of the cholera vaccine in the context of Indochina. The chapter ends with an examination of vaccine programs in Cambodia. These histories reveal the evolution of local choices on the ground and their influence on French medical programs. They are stories of the politics of scientific production and the science of policy implementation in the colony. The links and breakages between different parties are sometimes surprising, but they all speak to the priorities of stakeholders in public health and medicine in Cambodia.

## Vaccine Development and the Pasteur Institute

In the early years of colonial Indochina, the medical services regularly employed vaccines against four diseases: rabies, smallpox, cholera, and the plague. In the 1930s and 1940s a combined vaccine for typhoid and tetanus (TAB) and a tuberculosis vaccine (BCG)[10] came into occasional use. Vaccination seemed at this time to hold the promise of eradicating many of the major diseases of the world—although, except in the case of rabies and to a lesser extent diphtheria, no strong evidence of its prophylactic power existed, and except in the case of smallpox, no vaccine had proven consistently effective as a preventative.[11] However, these failures were not due to a lack of government interest or scientific research. For example, within French colonial Indochina, the vaccines for cholera and plague had a long and complicated development.[12] The smallpox and rabies vaccines, though developed elsewhere, also became closely intertwined with the history of science in Indochina. These vaccines were vital to the introduction of Western medicine into Cambodia, and their history is intimately tied to the research of the Indochinese Pasteur Institutes. Their development and clinical trials[13] were clouded in controversy, in Europe as well as its colonies. Even after many vaccines had come into wide employ, both their quality and supply were continuing issues in Indochina.[14] The Pasteur Institute, which was the only manufacturer of vaccines locally for most of the colonial period, came under fire repeatedly in Cambodia and other sectors of Indochina for its production difficulties.

Despite these problems, the French colonial government displayed an unwavering loyalty to the vaccination program. In order to understand this commitment, we will examine the events surrounding the research and inception of the cholera and plague vaccines—research that took place for the most part in Vietnam, Hong Kong, and India—before turning to the use of these vaccines in Cambodia. Along with highlighting the reasoning behind and the problems attending the use of vaccines in Cambodia, this analysis characterizes the relationship of the medical practitioners and researchers to the colonial endeavor. At the end of this chapter, we begin to sketch the response of the indigenous populations to this constellation of medical processes.

## Missionaries of the Pasteurian Order

If, as some historians of colonial science contend, medicine was simply a tool of empire, the vaccine was the weapon par excellence of penetration into local societies.[15] As we have seen in chapter 1, the mobile vaccinator was often the initial emissary of the conqueror into the conquered hinterlands—often not only the first but, for much of the colonial period, the only European many natives were to encounter. In Indochina, vaccination from the 1850s until the 1880s was done arm-to-arm, usually from child to group. A child would be vaccinated either from an attenuated animal vaccine product or from the arm of another child who had received attenuated vaccine in some form. This vaccination ideally induced a mild version of smallpox.[16] A pustule of the recovering vaccinated child was pricked by an alcohol-sterilized needle, and "vaccine" obtained from it was scraped in three rows, approximately thirty millimeters apart, on the arm of the recipient. Along with the scarcity of parents who were willing to allow their children to be donors, or *vaccinifères*, such a method created a risk of transferring other diseases with the vaccine (the authorities cited a risk of tuberculosis and leprosy, which were highly unlikely to be transferred in this way).[17] The practice also discouraged villagers from having their children vaccinated, since successful vaccination meant that they could be selected to accompany the doctor on the next leg of his trip. Despite some compensation for the inconvenience, the nature of the service was far from attractive and was less voluntary in practice than in theory. In 1890, the first vaccination rounds in Cambodia came to an end when locally produced vaccines ran out and no Khmers would volunteer as *vaccinifères*. Vague reports also suggested that the vaccines were not effectively attenuated. Immediately after the tour, a smallpox epidemic broke out in regions visited by the French vaccinator, and the indigenous population quickly refused to have anything to do with the vaccine. The depletion of vaccine supply was likely fortuitous for early efforts at public health, since it prevented further injury both to the location's population and to the reputation of French medicine. Despite these failures the health director was not discouraged, and he continued vaccinating Phnom Penh residents with locally produced vaccines (figure 2.1). He vaccinated 163 townsfolk in early 1891; all of these injections were reported failures.[18]

The difficulty and cost of getting an adequate quantity of fresh, uncontaminated vaccine from the metropole precluded wide-scale injection or scraping with animal vaccines. In an effort to improve the situation, in the late 1880s the GGI planned a fully equipped vaccination production facility in Indochina. To achieve this end, Undersecretary of Colonies Eugène Etienne[19] authorized a training period for Albert Calmette, a young graduate of the recently created École de Santé Navale in Bordeaux, at the new Pasteur Institute in Paris.[20] On finishing his supplementary training in 1890, Calmette's first assignment as both a Pasteurian and a *médecin général des troupes coloniales* was the organization of this bacteriology and vaccine institute. With the aid of the Pasteur Institute and the GGI, he quickly set up a laboratory on the grounds of the Saigon military hospital (figure 2.2).[21] When the first vaccines became available the following year, Calmette reported an increase in voluntary attendance at vaccination *séances* as well as improvement in vaccine effectiveness. During his brief tenure in Saigon, he continued to pattern himself after "*maître*" Pasteur, who was always mindful of the practical application of theoretical research. With an eye towards the economic and social needs of the colonial government, Calmette expanded his research into diarrhea treatments, alcohol fermentation, tuberculin, chol-

Figure 2.1. Phnom Penh Municipal Dispensary. Vaccination *séance*, circa 1890. Courtesy of the National Archives of Cambodia.

era, and snake antitoxin.[22] He did not enjoy the "disorder" and "inefficiency" of life in colonial Indochina, and he soon returned to France to continue his research career.[23] With his departure, another military doctor trained by the Pasteur Institute, Paul-Louis Simond, took over direction of the vaccine institute.[24] In the short term, the political situation and scientific exigencies in the colony were an ideal climate for the foundation of Calmette's work as a medical scientist. The institute he had organized in Saigon in turn became the foundation for the international reach of the Pasteurian research programs. His vaccine institute would later be known as the first overseas Pasteur Institute. Impressively, it had been founded only two years after the Parisian institute.[25]

Another key French researcher in Indochina was Alexandre Yersin. Yersin, like Calmette, was a graduate of Bordeaux and one of the first scientists trained at the Pasteur Institute. Unlike Calmette, he was driven to Indochina by the desire for adventure rather than ambition. He quit the institute in 1889, a year before Calmette was sent to Saigon. Yersin replaced a "drunkard" as ship's doctor with the French merchant marine company *Messageries Maritimes* and ended up in Saigon the following year.[26] In 1892

Figure 2.2. Saigon Military Hospital, circa 1890. Courtesy of the National Archives of Cambodia.

he organized an exploratory expedition into the Vietnamese highlands and eventually made his way overland to Phnom Penh. This would be his only trip to Cambodia in his fifty-year career in Indochina, although he would undertake two more mapping expeditions in the next few years, arranging to perform some token vaccinations in the highlands to supplement the funding for these trips. At the urging of Calmette, whom he had befriended at the Saigon Pasteur Institute in 1891, Yersin also signed on as a *médecin général des troupes coloniales* in 1893. He would shortly found a second Pasteur Institute, in the small seaside town of Nha Trang in central Vietnam.

The GGI intended that these laboratories be linked with the Hanoi École de Médecine, which was created in 1902 to train indigenous *médecins* for the anticipated expansion of the colonial health services to the native population.[27] Yersin, appointed in 1902 by Governor General Paul Doumer to organize and direct the École, was at first enthusiastic about this plan. Governor general since 1897, Doumer had displayed constant faith in science and public works projects to encourage the "development" of Indochina. Despite Doumer's pro-science stance, fellow Pasteurians Calmette and Simond were strongly against colonial proprietorship, urging the director of the Parisian Pasteur Institute, Emile Roux, to obtain ownership and control of the institutes. Calmette and Simond doubted that Yersin had the institutes' best interests in mind, and suspected him of subverting them for personal gain. They also strongly believed that only the metropolitan laboratory could direct an effective scientific research program in the colony.[28]

As would prove to be the case time and again, proximity was key to the implementation of policy. Yersin was the only prominent representative of the Pasteur Institutes on site. Doumer decided in discussion with Yersin that the labs were to remain the property of the Indochina government. Unfortunately for Yersin, at this key moment the climate for scientific research radically changed in Indochina, when Paul Beau replaced the pro-Pasteurian Doumer as governor general. Beau was a pragmatic, military-minded man who strongly disliked scientific elitism and thus, by extension, any special privileges accorded to the Pasteur Institute. In 1904, he appalled Yersin by placing the École de Médicine under the authority of the Health Corps and Inspector General Charles Grall. Even more intolerable, Beau also transferred administrative *and scientific* direction of all Indochina Pasteur Institutes to Grall. As head of native civil health services, Grall would likely have removed these laboratories from both military and Pasteur Institute direction. He most certainly would have removed them from Yersin's control. Thus, Yersin had a change of heart and decided that colonial government ownership of these laboratories was unadvisable. In a fury, he resigned from both the Nha Trang Pasteur Institute and the École. He worked fervently in the next two years with Roux and Calmette transferring ownership of the Saigon and the Nha Trang laboratories to the metropolitan Pasteur Institute, serving as contracted agents to the GGI. This was finally achieved in 1905.[29] However, the medical school remained under the control of the Health Corps and the newly formed colonial indigenous health services, the Assistance Médicale (AM). Thus the Pasteur Institute and the GGI essentially divided control of these two key sectors of the health system: the research laboratories and medical education.

Yersin and Calmette were to set the pattern for many employees of overseas Pasteur Institutes; they were military doctors assigned to special training at the Paris institute, then assigned *"hors cadre"* as Pasteur Institute employees in the colonies. At its founding, the doctors of the AM also came from this same pool of military doctors who were already present as part of the occupying troops in colonial territories. Similarly, the AM was intended as a public agency separate from military occupation. Thus, as it aged, it increasingly used nonmilitary doctors. With this shift, tension developed between military and civilian doctors, since the two groups had different promotion scales and often came from different social backgrounds and medical traditions.[30] These were turf wars of a strange sort. One could not accurately say that competition was fierce for open positions, since the government had constant problems filling vacancies in Indochina. For example, Calmette was to note, in one of the many negotiation sessions between the

Pasteur Institute and the GGI over control of the outlying laboratories, that contracts had to stipulate that the Paris institute would select and train all laboratory employees because "we cannot count on the Corps de Santé des Troupes Coloniales, which lacks doctors, to provide them."[31] Positions were available, but prestige was scarce. The perceived inferiority of civilian medicine on the part of the military doctors and the perceived unfair institutional advantage to military medicine seen by civilian doctors created political friction in the health services that lasted through the colonial period. This friction also affected the politics of the Pasteur Institutes.

## Controlling Science

The Pasteur Institute was often placed in direct competition with the AM on issues such as vaccine production and research priorities. This problem became exacerbated when the common denominator of military background was lost. For example, when civilian doctor Le Roy de Barres became the AM's local director of health in Tonkin in 1924, the contentious struggle for vaccine control became particularly fierce.[32] The AM believed that the cost of vaccines coming from the Pasteur Institute was too expensive, and that its own production and control of vaccines would be more cost-effective. It also desired some control over other laboratory services such as bacteriology and potable water analysis. In contrast, the Pasteurians believed that their specialized expertise made them the most (indeed the only) qualified group in the colony to undertake cutting-edge scientific research and to oversee a process as technically complicated as vaccine production. The argument regarding sufficient expertise was not unreasonable; at the turn of the century, only a handful of medical schools in the world offered a level of training equal to that of the Pasteur Institute in new laboratory-based medical fields such as bacteriology and immunology.

A further problem with the health services was the division in their chains of command. As mentioned in chapter 1, the inspector general of hygiene and medical services of Indochina was responsible for overseeing the health services in Indochina's five regions. However, the local directors of health were hired and fired by the *governor general*. Technically responsible for setting the tone of all the medical services in Indochina, the inspector general in reality served in an advisory role to the governor general. Thus, while in theory the inspector general outranked and oversaw the local health services, in actuality each local director of health answered to the civil administration before the head of the health administration. This strange schism in responsibility created conflicts in medical, scientific, and social priorities.

A prime example of such a conflict is the history of the bacteriology laboratory in Cambodia. In the early decades of the twentieth century, the French administration in Cambodia had expressed an interest in building a Pasteurian laboratory facility. In 1910 the RSC budgeted the funds and requested an expert to organize such a laboratory. The inspector general arranged with the Pasteur Institute to send a trainee to complete this task. Noël Bernard, a young military doctor recently recruited by Calmette, was sent from Paris Pasteur Institute training to Indochina for this purpose. Upon his arrival in Saigon, he was temporarily diverted by the governor general to Hué, in central Vietnam, to study the urgent problem of potable water. (This was not the first time a Pasteurian assigned to Cambodia had been thus diverted; a year earlier, the Pasteurian Denier had been sent for the same purpose, had been reallocated by the RSC, and had spent the duration of his assignment studying sick cattle.[33]) The résident supérieur of Annam and the general director of public works had originated the request for Bernard s temporary diversion to Hué, and the governor general then confirmed the changed assignment with the Parisian Pasteur Institute, but the man in charge of health services, the inspector general, neither advised on nor approved the changed mission. The governor general also failed to consult the Cambodian colonial government before the diversion. The governor general consistently subordinated the needs of the medical service to those of the civil administration, and in this case the needs of the Cambodian medical service were subordinated to those of Vietnam. When Paul Louis Simond, another Pasteurian who had served as director of the Saigon Pasteur Institute from 1896 to 1898, was to be moved from his teaching position at Le Pharo[34] and assigned as the inspector general in Indochina in 1910, he wrote in anger to his mentor Calmette,

> M. Grall is certainly good to reserve one of the posts of Indochina for me! Those posts have nothing for officers of my grade, neither initiative, nor authority, nor influence over hygiene or anything involving hygiene, nor any sort of material advantage. . . . If I go there, it is only because I cannot see any other way to earn a salary necessary for a decent living, however modest. It would be neither to fulfill an inexistent obligation nor to provide useful service.[35]

Simond would end up accepting this posting during World War I, from 1914 to 1916. However, many of the conflicts that he foresaw in 1910 would lead to his complete resignation from the Health Corps in 1916.

Whatever the reasoning behind Bernard's 1910 diversion to Hué, his temporary assignment stretched out to twenty-five months. By the end of these two years, he had used the funds and equipment originally designated for Cambodia to organize a Hué bacteriological lab. Cambodia would make do with its rudimentary laboratory until 1915.[36] The laboratory Bernard established in Hué quickly became the center of a particularly acrimonious debate between Yersin as representative of the Indochina Pasteur Institutes and the military Health Corps in Indochina, both of which demanded ownership and control of the facility. In May 1912, Governor General Albert Sarraut urged both parties to leave matters as they stood, with the Pasteur Institute having majority control over the institution since it owned all the equipment while the GGI owned the building. By August, the director of the Parisian Pasteur Institute signed over title of the equipment to the Health Corps in a gesture of goodwill, although Pasteurians continued to staff the laboratory.[37] In 1923, during the renewal of the business contract between the Indochina laboratories and the GGI, the ownership and control of the Hué lab were transferred back to the Pasteur Institute.[38]

Even given their training, these overseas Pasteurians were not medically infallible. In fact, due perhaps to their greater faith in the promise of experimental medicine and their frequent activity in experimentation (with human subjects), they were often embroiled in significant research controversies. With few exceptions, these controversies within the medical administration remained hidden from public view. Whatever internal frictions existed, the GGI, the Health Corps, and the Pasteur Institute repeatedly cooperated in hiding each other's errors from the public. We will focus on three such situations in detail: Yersin's plague research, cholera vaccine experimentation on indigenous soldiers, and a series of tetanus deaths linked to the cholera vaccination. These episodes reveal the political and professional context of the medical services, as well as the historical contingencies that affected their development. They also serve to situate the colonial medical service in relation to both local realities and international medical trends.

## Plague Research

Upon descending into Saigon in May 1894 after a particularly exciting expedition in the highlands of Vietnam, Yersin found waiting for him a directive to report to Yunnan in southern China, where he was to study the ongoing plague epidemic. The governor general had become increasingly worried by the spread of the bubonic plague through parts of China, India, and island Southeast Asia,[39] and had authorized a trip to research ways to prevent and treat the disease. Yersin, stopping at Hanoi on the way to Yunnan, convinced the interim governor general to send him instead to

Hong Kong, the epidemic's center. Arriving in Hong Kong a few days after the Japanese research contingent headed by Kitasato Shibasaburo, Yersin found himself out of favor with the English colonial government.[40] On the day of Yersin's arrival, Kitasato had published a report claiming that he had discovered the plague bacillus. However, discrepancies in the report left Yersin unconvinced as he launched into his own research program.[41] In the next few months he identified and published a separate report identifying the plague bacillus, and within two years he developed and tested a serum against the disease. Problems with Kitasato's report, including an inaccurate morphological description of the bacillus, persuaded the international scientific community to assign Yersin priority of discovery.[42]

In the initial clinical tests in Canton in 1896, Yersin reported that his serum decreased plague mortality rates from over 80 percent to 10 percent. These proclaimed successes led to much internal discussion among French officials about the possibility of creating a Pasteur Institute in Canton. These discussions explicitly linked the presence of such a laboratory with increased French influence on mainland China, in watchful proximity to the English colony. In the end, the hostility of the Chinese and English militated against the immediate creation of such a facility. However, Yersin's successes with his serum were convincing enough that by November the British government sent a preliminary inquiry to the Indochina government about the cost and quantity of plague serum that it might be able to provide to the British colonies.[43]

In a confidential letter to the governor general, the French consulate in Hong Kong advised that any response should consider Britain's lack of respect for Yersin's achievements. To provide the serum to English doctors who were systematically hostile to French science and so ignorant that they could not recognize the microbe would result in the incorrect employment of the remedy, which would then drag down the reputation of French science. The consulate proffered the following excuse for French refusal: "It would be imprudent and incompatible with our national dignity to be placed in the position as sellers of a medication whose efficacy is not recognized by many prominent doctors, and so in consequence, we find ourselves constrained by the current state of the science to reserve the use of Yersin's serum exclusively to its inventor and to the doctors in whom he would consent to entrust its use."[44] It seems that the GGI did not enter into negotiations with the British government on providing the serum, although it prudently also did not provide the consulate's explanation. It is unclear whether the British lost interest or the GGI simply did not respond. Britain had more than a passing desire to find an effective treatment for the plague. Not only Hong

Kong but also its largest colony of India had continual problems with the murderous and economically devastating disease.

Thus, the cold shoulder of the French did not deter the British. When a plague epidemic exploded in India in early 1897, the British Raj funneled its scientific and financial support to another well-known researcher, Waldemar Mordechai Haffkine. Of Ukrainian origin, Haffkine had worked briefly at the Pasteur Institute, in fact replacing Yersin as the laboratory *preparateur* in 1890. After his stint at the Institute, he quickly associated himself with the British research establishment through his work on cholera. Shortly after Yersin, he had developed another version of the plague serum.[45] Along with Yersin and Haffkine, the British Raj invited other leading international microbiologists to India to study the disease, irrespective of nationality.[46] Yersin at first was disinclined to accept the invitation; however, upon hearing of Haffkine's plague work in India, Yersin soon departed Nha Trang for Calcutta.[47] Throughout 1897 both contingents performed thousands of vaccinations on Indian populations, testing their sera as both prophylactic and cure. The results of their experiments were more than a little disappointing. In terms of prevention, Yersin's serum lasted a couple of weeks at most. Haffkine's seemed to work somewhat longer, but patient reactions were often severe. By the end of the year, Haffkine claimed slightly greater success than Yersin in both prevention and cure. Yersin's serum showed, with the most optimistic reading of the data, a 33-percent cure rate—only a slight improvement over the untreated rate of 20 percent. Discouraged, he had returned during the summer to Nha Trang to work on perfecting the serum, leaving fellow Pasteurian Simond in India continuing the clinical trials (figure 2.3).[48]

Although the experimental results were disappointing, colonial governments had few other options, and continued to employ these sera. In Indochina the Yersin serum was faithfully used for the duration of the AM, although by 1908 the service tentatively implemented the Haffkine serum (an episode that we will examine at the end of this chapter). Having sent his serum to his laboratory in Nha Trang, Yersin would claim for the next few decades to be experimenting on improvements to it, although no records exist that he modified it any further. His lab became the center for production of the plague serum for Indochina, and the GGI would continue to support this lab despite some serious mishaps with the bacillus.

The first controversy began shortly after Yersin's return to Nha Trang from India. Before 1898, no official plague cases had been registered within the borders of French Indochina. In June 1898, however, several cases were

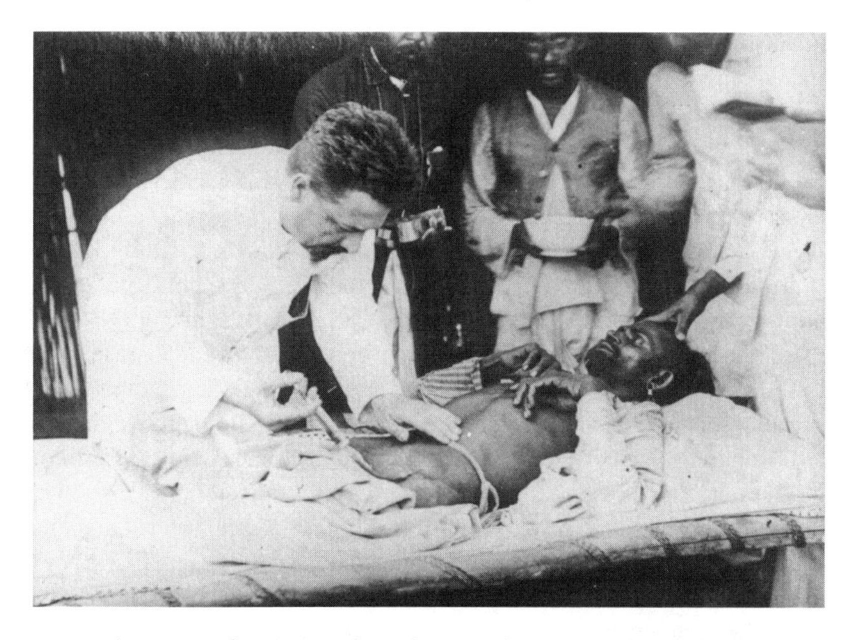

Figure 2.3. Paul-Louis Simond inoculating pestilent man in India, circa 1898.

declared in the homes that were in closest proximity to Yersin's Nha Trang laboratory. The first death occurred on June 23, 1898, as Yersin was en route to catch a steamer to Hong Kong for yet another plague epidemic. Upon hearing the news, he immediately returned to Nha Trang and instituted preventative vaccinations among the villagers closest to the laboratory. Through June and July he inoculated more than four hundred residents.[49] Then he waited anxiously to see if any more cases occurred. When a young girl died on July 22, Yersin was to note that her swollen buboes disappeared before her death. Although Yersin had devoted his entire career to basic research rather than medical practice, he interestingly reverted to giving priority to what was essentially a clinical diagnosis over a bacteriological analysis. Judging the victim by the appearance of her buboes, and using no laboratory analysis, he opined that "perhaps" plague was not the cause of her death.[50] Even more troublesome to the doctor, a woman he had vaccinated on the sixteenth of the month had developed the plague by July 26. Yersin developed another unlikely theory for her affliction, arguing, "I think that before she had been vaccinated, she was already incubating the disease. The preventative serum being too weak was only able to abnormally prolong the duration of incubation."[51] He injected her again after she became ill, but the treatment had no "beneficial effect." She soon died. By mid-August more than twenty cases had been registered, with fifteen deaths. To complicate matters, a cholera epidemic had also erupted amongst the population. At this time, measures against cholera—involving disinfection and regulations for burial of the dead—were milder than plague measures. Because incineration of huts had become standard procedure with the plague, villagers regularly misrepresented plague deaths as cholera to French authorities.[52]

Toward the end of August, with no end to the epidemic in sight, Yersin wrote a letter to his mentor Roux revising his view of the epidemic. Whereas in earlier correspondence he took as a given the leak of the disease from the lab to the outside environment,[53] by mid-August he changed his interpretation of events. On August 18, he had heard of an unidentified disease afflicting the village of Binh Dinh one hundred kilometers to the north of Nha Trang. This disease had killed thousands in five regional villages over the past four years. With little supporting evidence, Yersin decided that this was the origin of the local plague epidemic. He urged the GGI to investigate, but after receiving no response, he sent his assistant Carre upriver to investigate. By September, Carre had reported back that the disease was unlikely to be the plague. Upon hearing the news, Yersin reported to his mentor that he was awaiting a "full report" before he could reach any final conclusions. However, Carre's investigation was completed and he sent no further information. He left the region shortly after this report to study livestock diseases in Cambodia. In the meantime, Yersin had organized a quarantine village near the original village and stepped up the burning and disinfection of local homes.

Governor General Doumer was displeased to hear the news of the first confirmed plague epidemic in Indochina erupting at the very site created for its prevention and control. Yersin noted with some petulance that Doumer had no grounds to be angry with him, since he had sent Yersin to organize a hill station[54] in the Lang Bian mountains (in Dalat, central Vietnam) in March through May, the interval during which the plague had probably leaked out. Having thus been called away by the governor general himself, Yersin could not be responsible for mishaps in his unsupervised lab. By October, Yersin was in a state of high defensiveness. He complained of the journals that "howled" blame at him, and he wrote to Roux, "I defy that anyone can prove that the plague originated from the laboratory."[55] A colonist threatened to sue the Pasteur Institutes for damages and interest in October. To improve the situation for the medical services, the AM medical chief wrote a letter on Yersin's behalf to the governor general, supporting

Yersin's theory that the disease had originated from Binh Dinh, yet failing to mention (and perhaps not knowing) that Yersin's staff had already discounted that theory. Inspector General Alexandre-Marie de Kermorgant wrote a similar letter to Roux in Paris, arguing that the plague must have originated from elsewhere since no one from *within* the laboratory had been infected, only those *surrounding* it. Both the medical researchers and civil administration of Indochina would embrace this dubious epidemiological reasoning. At year's end the epidemic had subsided, and the GGI, the Pasteur Institutes, and the AM all aligned behind the theory that the plague had arrived in Nha Trang, skipping all major ports on the way, via a Chinese commercial junk. Yersin would feel sufficiently confident by December to note with some exasperation that the new *résident* of Annam was a "nervous, unscientific" fellow who was unable to accept the notion that the plague had originated from elsewhere.[56]

Sporadic plague epidemics occurred from 1898 onward in various parts of Indochina, and the rumors of its Nha Trang origin would continue. In 1903, when the epidemic hit Hanoi, the newspaper *Le Petit Tonkinois* ran a commentary on a rumor that the prominent civilian doctor Le Roy des Barres had brought it into the city. The journalist argued that the rumor could only have been spread by "those who have some interest in raising their prestige, singularly at a low in the colony, the doctors of the health corps." He suggested that the most likely candidate for importing the malady was a person in this category: Yersin.[57] Again, there was a split between civilian and the military factions, but also between research and practice. Accidents in later years supported the lab's culpability as a source of the plague. In 1921 the plague broke out in Nha Trang immediately after the death of a number of visibly pestilent rats along the verandah of the laboratory. Yersin became angry with local administrator Lagrange, whom he complained was playing the blame game. One would assume that the Nha Trang Institute improved security against biological leaks after these problems. However, if Yersin did take precautions, they were not effective. In the following year, 1922, one of the indigenous lab staff members who had injected human plague into horses died in a suspiciously plague-like manner. The dead boy, Vinh Tham, was a son of the former king of Annam, Thau Thay, who was then exiled in Réunion.[58] No investigation ensued and little publicity was generated. Despite these additional strongly suggestive incidents, it was the story of the disease coming to Vietnam via Chinese junk, a story reproduced continually by the three wings of the medical service, that would endure into modern times.[59]

## Experimenting on Conscripts

The history of the cholera vaccine in Indochina repeats many of the themes of the story of the plague vaccine, although its first use in Indochina occurs a quarter of a century later, with the outbreak of the Great War. In 1916, during the height of World War I, the French government ordered the GGI to recruit twenty thousand infantrymen and thirty thousand workers for the European front. Collected from the most impoverished peasants of Vietnam, Cambodia, and Laos, many of these men were in poor physical condition. Crowded by the thousands at departure ports such as Haiphong and Hanoi, they suffered from periodic outbreaks of diseases such as scabies, tuberculosis, and syphilis. The most troublesome disease for the Indochinese government was cholera. In February 1916 the Ministry of Colonies alerted the governor general of cholera deaths on ships transporting Vietnamese battalions to Europe. Unlike other debilitating health problems that also drew complaints from the metropole, the threat of cholera was significant enough that the ministry was willing to turn back entire steamers contaminated with the disease. To solve the problem, Inspector General Paul Louis Simond authorized research on a cholera vaccine.

Multiple cholera vaccines had already been developed in the previous century. Robert Koch (of Koch's postulates fame), the Pasteur Institute, Haffkine, the German team of Ludwig Brieger and August Paul von Wasserman, and the Spanish researcher Jaime Ferran all claimed to have produced a cholera vaccine by 1895. Technically, Ferran performed the first experimental cholera vaccinations; however, because the vaccine he employed was not "fixed" (attenuated by passage through several generations of a laboratory animal), his peers viewed his work as sloppy, hazardous, and unscientific. Haffkine ultimately received the most administrative and scientific support for his claims to priority. Although Haffkine was to gain considerable prestige in the last decade of the nineteenth century with his cholera vaccine, several well-publicized tragedies surrounding both his cholera and plague vaccines in India led to his marginalization by the scientific community and a forced semiretirement by 1905.[60] In Indochina, Denier had experimented briefly and unsuccessfully with a cholera vaccine in 1906. Despite these various efforts, no party had developed a truly effective and safe cholera vaccine by World War I.

Simond assigned Dr. Alexandre Gauducheau, the director of the Institut Vaccinogène and Laboratoire du Bacteriologie de Hanoi,[61] the task of producing a cholera vaccine for Indochina. It is likely that Gauducheau based his vaccine on the research of Denier a decade earlier. He began trials of his newly manufactured vaccine on the inmates of the protectorate prison in Hanoi in March 1916.

The prison, which conveniently was suffering from a cholera epidemic at the time,[62] had approximately 317 inmates. Participation in Gauducheau's experiment was theoretically voluntary, although coercion was probably used. Sixty-two prisoners refused vaccination and 253 "volunteers" participated. Gauducheau found a drop in cases from 6.3 percent of the unvaccinated prison population to 4.8 percent of the vaccinated group. The vaccine decreased, but did not eliminate, the occurrence of cholera. Mortality meanwhile was significantly reduced from 4.8 percent among the unvaccinated to 0.7 percent of the vaccinated. Gauducheau would note that this mortality reduction may have been due in part to a changed—one can assume improved—regimen for prisoners who willingly accepted vaccination.[63] The results were sufficiently encouraging so that the Hanoi Institut Vaccinogène and the Nha Trang and Saigon Institutes were producing the cholera vaccine by early May, even though officials admitted that the results were not yet conclusive.[64]

By May 5, indigenous conscripts departing Hanoi and Haiphong to the European front were being vaccinated en masse. The cholera journals of the vaccinating doctors reveal that sporadic cases occurred amongst the vaccinated contingents. In one of the most disappointing series, 1,300 recruits stationed at the port of Kien-An were vaccinated from March 29 to June 29, with sixty-four "serious" cases of cholera occurring among the vaccinated.[65] Given these results, it was not surprising that outbreaks continued to occur among battalions that left for Marseilles. Indeed, the problem of cholera seemed to increase significantly in May, concurrent with the increased vaccinations, for more ships were refused at ports along the way to Europe. Confusion reigned as to the origin of these cholera epidemics. By May 10 the minister of colonies, Gaston Doumergue, ordered Governor General Ernest Roume to suspend contingents departing from all points in Indochina. Roume urged the metropole to reconsider; leaving thousands of men crowded in makeshift camps in Vietnamese ports would be both dangerous and disruptive. He assured Doumergue that the Indochinese colonial government would take all necessary precautions and *increase* vaccinations. Although on May 16 the metropole repeated the order of suspension, on May 19 Doumergue allowed Roume to send limited numbers to the European front, provided that the Indochinese government undertook all necessary precautions. What these precautions were to be is unclear. Cholera vaccination ended with the cessation of the epidemic in July. The medical authorities dared not credit the vaccine with the end of the epidemic, noting only that they suspended vaccinations when new cholera cases ceased to appear.

The severity of the May epidemic had political repercussions. By the year's close, the Paris government demanded a report on the origin of the Indochina cholera vaccine from the minister of colonies. Oddly, the responsibility for this study fell to Gauducheau. In his investigation of *his own* vaccination research, he pointed out that he had "prudently" decided to inoculate some prisoners before the sixty thousand conscripts. Taking some liberty with his data, he claimed that the "perfect" results of these prison vaccinations justified its use on all Annamites shipped to France. He confidently concluded that there had been no negligence on the part of the sanitary authorities of the colonies in employing the vaccine. If anything, these vaccinations represented "the scientific perfection and methodic application of a sound prophylactic."[66] Despite the deaths of hundreds of men linked to these vaccines, the research went no further. Interestingly, Gauducheau had at hand all of the research tools of his European counterparts, including access to sophisticated statistical techniques that he chose not to use.[67] His weak results were due not to lack of knowledge or equipment, but rather to poor experimental design and less stringent epidemiological reasoning. Certainly, such poorly conducted experiments were also occurring in the metropole. However, those experimental conclusions would not likely have stood up to peer critique if validation had resulted in action on Europeans or Americans.

This episode indicates not just that the validity of science to authorize certain actions is highly dependent on sociopolitical context—a well-established truism of science studies—but also that epidemiological reasoning in the colonies (and perhaps scientific reasoning generally) was less stringent because its application had fewer implications for the politically enfranchised. The subjects of the medical research were socially and politically disconnected from those conducting the research. Another commonly held belief about colonial science—that science in the "periphery" lagged behind metropolitan science because colonial researchers were themselves somewhat professionally disenfranchised from the centers of science—is somewhat incorrect. Gauducheau had available the methods and resources to produce "cutting-edge" research in the colony, but sociopolitical context (not scientific context) may have discouraged the highest standards of epidemiological reasoning. The problems with the cholera vaccine had not been resolved, and seven years later this would become the center of one of the most serious challenges to the Pasteur Institute. In the next section, we will examine the incident in detail. Despite these issues, the vaccine itself would continue to be used in Indochina through the 1920s and 1930s.

## Dirty Needles or Doubtful Science?

On June 5, 1922, a Vietnamese *infirmier* vaccinated a family of eight against cholera in an indigenous home in Sadec, a town southwest of Saigon, after a choleric died in the neighboring home. By June 10, the entire family had developed tetanus.[68] Nöel Bernard, then director of the Saigon Pasteur Institute, soon launched an investigation with the director of health services for Cochinchina. By the end of the month, six of the eight vaccinated had died of tetanus. The negative publicity in the local newspapers and acrimony from the Sadec community was intense. Both the indigenous medical personnel and the French doctors blamed the Saigon laboratory. Nervous doctors in Sadec reported putrid odors coming from other vaccine flacons. To relieve public anxiety, Comte and Bernard arranged to take a batch of flacons from Sadec and open them in the presence of neutral witnesses and journalists in Comte's office.[69] Laboratory inspection of these flacons revealed that at least three had been previously opened and reclosed in Sadec; they were found on laboratory analysis to contain various microbes. Two others that were still factory-sealed and opened for the first time in Comte's office had no microbes. Bernard concluded that the Sadec medical staff was incorrectly resealing and reusing opened flacons of vaccine. He assured local doctors that the putrid odor was simply the phenic acid in the solution, and that the rubber stoppers on the flacons were causing any sulfurous smells. If doctors had any doubts about the quality of the product, they were to convey the vaccines back to the laboratories.

However, Bernard refused to perform a general bulk recall of the cholera vaccine from area medical services, noting that he was "absolutely certain" of the quality of the vaccines. The Saigon lab had produced 209,952 cholera vaccination doses, in 13,484 flacons, in this year. He reasoned that if the vaccine had been contaminated at any point in the manufacturing process, it would have occurred in more than one of 13,484 flacons. These Sadec deaths were the first reported incident of tetanus, and thus probability demanded that the fault must have been with the handling of the vaccine. In his June 16 report, Bernard listed four possible sources of the tetanus: (1) contamination during vaccination preparation at the Pasteur Institute; (2) contamination during the course of vaccination by the "Annamite *infirmier*"; (3) improper use of a vaccine flacon that had been previously opened, contaminated, and then reclosed; and (4) latent tetanus in the vaccinated persons that manifested itself after the cholera vaccination. Bernard spent the bulk of his report eliminating the possibility of contamination during manufacture (option 1), and a paragraph eliminating the possibility of option 4, "latent tetanus." Options 2 and 3, he believed, were the most likely. He did note that witnesses confirmed that the "Annamite *infirmier*" had correctly flame-sterilized the needle and performed the vaccination. In each instance that he mentioned the act of vaccination, Bernard specified that the *infirmier* was "Annamite" and had been unsupervised by a doctor. He also noted that, no matter how conscientious the *infirmier* may have been, aseptic conditions would have been nearly impossible to maintain in an indigenous home.[70] In repeatedly suggesting a connection between the contamination and the racial status of the victims and the medical practitioner, he reinforced the racial and technical inferiority of this group in relation to European supervisors and the Pasteur Institute. Bernard would end his report with an elegant argument for minimizing this local tragedy, and for the counterproductiveness of placing blame on the Pasteur Institute.

> In the particular case of Sadec, is it legitimate to interpret in favor of the vaccinator, a simple *infirmier*, all the uncertainties resulting from his insufficient surveillance, all the exceptional conditions of the accident, and instead to throw the responsibility on the organization that has employed all necessary safeguards in the past and the present? [¶] Is it appropriate to cast suspicion on the use of all vaccines among the indigenous population, making them fear the possibility of a contamination that is able to bypass the strictest controls, in order to defend an *infirmier* whom no qualified witness saw act and whom no one dreams of punishing? [¶] Is it not preferable that, while genuinely sympathizing with the completely legitimate local emotions, we use all of our force of persuasion to contain, limit, and channel this fear, to prevent the alienation of the population of this country from prophylactic methods that provide the greatest benefits against the epidemic diseases that constantly threaten them?[71]

This argument encapsulates many of the issues at stake in the vaccination development programs of Indochina. In the colonial context, the ability to legislate local behavior may have been substantial, but this power often did not translate to actual changed behavior. The ability to improve health via changes in infrastructure (clean water or sewers, for example) was economically unviable. Further, the AM constantly suffered a shortage of medical staff. Vaccines held the greatest promise of improving indigenous health at the lowest economic cost. Further, in the days before "miracle" medicines such as sulfa drugs and antibiotics, vaccines were often the only defense against many of the acute diseases present in the tropics. If we consider the four main diseases against which vaccines were employed—plague, cholera, smallpox, and rabies—no medical technology except vaccination was effective in either their prevention or their treatment.[72]

The month following the Sadec deaths again saw the revision and the promulgation of an official sanitized interpretation of events. On June 22, 1922, both the military medical service and the AM staff received a letter from the director of health services in Cochinchina and Cambodia informing them that the tetanus deaths were a localized and exceptional incident. The letter placed no blame either on the *infirmier* or on the laboratory. The director of health services sent a similar letter to the GGI, explicitly stating that neither the Pasteur Institute nor the *infirmier* were at fault. The letter that the director sent to the inspector general shortly afterwards noted that the local medical chief and the Pasteur Institute were unable to determine who was at fault, and that in any case it was a localized event. And again, the vaccine continued to be used.

All the above can be seen as examples of the collusion of the colonizer against the colonized, appropriation of the colonial body for biomedical research by Western researchers, and the use of the colony as a laboratory for improvement of metropolitan science. While these readings are partially valid, the processes of negotiation and control involved more than the dynamics between colonizer and colonized. In some instances, the arrogance of the scientist over his subject was a greater impetus to these researches than the arrogance of the imperialist over his subjugated natives. Sometimes the two justifications for domination overlapped. Many other scandals within the medical establishment occurred among the French population in the colony, a few of which we have mentioned in passing; these incidents, although not involving colonizer and colonized, reveal the same patterns within the medical services. The patterns of experimentation, error, and collusion would indicate that control went beyond the domination of the native. At stake was not only the image of the colonial state, but also the autonomy and authority of the medical profession. This is illustrated in the following vignette involving a young French lawyer in Phnom Penh.

## Questioning Medical Authority

A French lawyer in Phnom Penh, Guy Tromeur, hospitalized his young daughter Marie-Louise on July 16, 1931. She had been diagnosed by Health Director Joseph Victor-Eugene Bouvaist with amoebic dysentery on July 9 and treated at home for several days. Although Bouvaist believed she was no longer in danger, he recommended that she be hospitalized for a few days of observation. At the hospital, Marie-Louise's attending physician was Dr. Pierre Escale. Without consulting with her former doctors, Escale radically revised the child's medical regime from small doses of emetine to electrargol, an aggressive chemical treatment, and daily routines of *draps mouillés* (wet sheets). Her health began to decline, and her condition over the next week fluctuated. On August 1 she was still "febrile" but conscious, talking, and smiling. On that day, Escale injected her with seven grams of the drug emetine. Three hours later she became extremely agitated and unable to speak, and by the next day she died. An autopsy listed her cause of death as the combination of amoebic dysentery and abscess of the liver.

Tromeur demanded that the *résident supérieur* of Cambodia, Fernand Lavit, investigate the doctor. Lavit sent an inquiry to Health Director Bouvaist, who replied on August 24 with a dense explanatory missive sprinkled with many scientific details about colibacillosis and current options for its treatment, quoting freely from a leading French medical textbook. The reason why Bouvaist discussed colibacillosis was unclear, since it was a diarrheal disease distinct from amoebic dysentery, Marie-Louise Tromeur's official cause of death. This may have contributed to further suspicion on the part of Guy Tromeur. After discussing colibacillosis, Bouvaist returned to amoebic dysentery, arguing that the medical arsenal for dysentery treatment was greatly limited and that emetine was considered among experts to be the best choice, even taking into account its admitted toxicity. Further, Bouvaist included a strong defense of his beleaguered medical staff, a defense that tacitly admitted mitigated responsibility for negligence: "Mr. Tromeur could not know or observe the effort requested of all the European personnel at the Hospital and the personnel at the Maternity, an effort accepted by all with a rare abnegation. . . . If one reflects that in this hospital, which treats nearly 500 patients a day, a full night's sleep for a doctor is exceptional, it seems difficult to say that Mr. Tromeur's child was not cared for with all possible devotion."[73]

On September 16, 1931, Tromeur sent a reply to Bouvaist via Lavit. Unintimidated by Bouvaist's scientific language, he had researched the textbook quoted by the health director and discovered several Parisian medical critics of that same book. These critics held the view that no one actually died of colibacillosis. Tromeur had also read further in that same book of other recommended treatments that Escale had not followed. The young lawyer demanded a homicide investigation.

Lavit was hesitant to take the matter any further, so Tromeur appealed to the next level of medical authority. He sent a request to the inspector general of Indochina, who opened an inquest. By December of 1931, the inquest had been decided in favor of Escale. Tromeur demanded a copy of the inquest findings, but in mid-February the GGI refused his request. In March 1932, Tromeur filed a judicial complaint and turned to the newspapers, writing a letter to *La Presse Indochinoise* in which he explained, "I have never tried to conceal my repugnance for judiciary action, and would not have considered it if it were not that one observes in Indochina that a doctor may kill without causing the slightest ripple."[74] Despite the public attention, the health director refused to give Marie-Louise's hospital files to the judiciary, citing medical confidentiality. In April 1932 Escale wrote a long memo in his own defense, which was attached to this refusal. Tromeur then contacted the Ministry of Colonies in Paris. Finally, in September 1932, over a year after Marie-Louise's death, orders from the metropole authorized the release of her files. By this point Escale had gone on medical leave, replaced by Dr. Antoine Fabry, who was also escaping his own scandal.[75] On the morning of September 30, 1932, the local head of the Sureté, the *juge d'instruction*, and Tromeur met Fabry and Bouvaist at Escale's former office in the hospital to retrieve Marie-Louise's file. It was not there. The health director speculated that it "was misplaced"—"probably during Escale's departure." Having exhausted all his options, the affair ended in bitterness for Tromeur.[76]

Even an intelligent, educated, and determined lawyer, fully vested with all the rights of a Frenchman, was unable to bring the acts of the medical establishment into question. The closing of ranks amongst the health staff was a collusion of the doctors against a patient as much as of the colonists against a native. It was also part of the wider struggle within the scientific and medical professions of the West to maintain professional prestige, benefits, and the rights to self-regulation. Much has been written about the processes of medical and scientific professionalization in the metropole.[77] Some studies have at least mentioned similar processes in the colony.[78] Few if any studies relate the two subjects. When one situates colonial medical doctors within the story of the medical professionalization movement in the West, as well as among the lower echelons of the colonial administrative apparatus, their motivations become much more complicated. In the metropole and the colony, the actions of the doctor were still assailable by the layperson. It is also likely that the clinical staff (as opposed to the research staff), at a remove from metropolitan centers, may not have been completely up to date on their medical knowledge. Nonetheless, Escale and his staff used the growing professional arsenal of the doctor (i.e., medical confidentiality, specialized knowledge) to shield themselves from external blame. Further, the colonial government, unlike the metropolitan government, had a vested interest in quashing intra-French disputes. French prestige in the eyes of the "native" was of constant concern to colonial administrators. However, we cannot simply subtract the needs of the state and its agents from the needs of citizens and colonial subjects. Even colonial administrators could be frustrated by the murky priorities of the medical staff, as the following vignette reveals.

## The Cambodian Vaccination Services

During a severe cholera epidemic in 1926, the *résident* of the northeastern Cambodian province of Stung Treng, Julien Mercier, wrote an agitated letter to the *résident supérieur* complaining of the medical service. He had witnessed the doctor vaccinating a number of healthy men rather than those already sick with cholera. He attributed this choice to a shortage of the vaccine, writing, "I do not know how much a dose of the cholera vaccine costs, but I prefer to incur enormous . . . costs rather than see a brave man perish before my eyes. This is what I experienced earlier. . . ."[79] The RSC, as was often the case, found himself the middleman between the provincial government and the health service. He forwarded the letter to Dr. Bernard Menaut, then temporarily serving as the health director. Menaut answered with a short and mildly patronizing letter, requesting that the RSC would, in his communications with Mercier, "[please] ask him not to waste the vaccine sent to him by using it as an anticholera medicine, as this vaccine has no curative properties and its use on those already infected or suspected to be infected may give rise to accidents."[80] Mercier, it seems, mistakenly believed the cholera vaccine to be both a preventative and a cure. This is not surprising, since the vaccine had been used in both ways shortly before this episode. The medical service, however, had expended little effort in explaining the change to the public. Indeed, Menaut's brief missive was the only response by the medical service to the administrator's confusion. The brevity of explanation could be attributed to the overwhelming demands on the medical service during this period of epidemic, or perhaps to a perceived lack of need for lay folk to understand the distinction between a preventative and a treatment, or even to an effort at being lighthanded and diplomatic during a time of stress. However, this lack of communication was not simply between scientists and bureaucrats, or between doctors and patients. It often reflected a lack of clear knowledge. As we have seen in the previous example, perhaps the medical staff could not diagnose with certainty whether Marie-Louise Tromeur had colibacillosis or amoebic dysentery, but they were compelled to act as if they were certain. On a more general level, while experimental vaccines were proliferating in the first quarter of the twentieth century, even specialists did not understand exactly how they worked at the immunological level.

In Cambodia, unreliable vaccines were a theme throughout the colonial period. Khmers first experienced the cholera vaccine after World War I, shortly after the research of Gauducheau. The first reference to its use in Cambodia is in the annual report of Dr. Honoré-Mathurin Le Nestour in Kompong Cham, who observed in 1919 that the vaccine was "placed at the disposition of the local population" but did not find much favor, since the natives already used a local remedy called the "Khmer Mixture."[81] The vaccine was not widely used in Cambodia until a severe cholera epidemic of 1926. Seemingly effective at first, it quickly became suspect. By the end of the year, the chief of the newly created Hygiene Service observed that

> [t]he [cholera] vaccine, which could have been a precious aid as it has been on several occasions in the past, was revealed to be ineffective so much so that one could describe epidemics of the vaccinated in Cambodia. [¶] This vaccine deficiency must be attributed to either a problem in its preparation or to the variety of germ that was rampant this year being of a new origin and distinct from the vaccine stock.[82]

As mentioned earlier, the unreliable nature of the cholera vaccine had been with the product from its inception in Indochina. In 1928, the medical chief of Kampot would mildly note that the cholera vaccine was "not very helpful . . . . In every case up to now, it seems that in the provinces it has not produced the results that were expected."[83] The latter half of 1928 marked the beginning of vigorous patient aversion to the cholera vaccination, particularly in Kampot, Prey Veng, and Battambang. Both Khmers and Europeans refused to be vaccinated. The Health Director identified the problem as one of vaccine quality, noting that several prisoners contracted cholera immediately after vaccination from a particularly cloudy flacon. He also observed that some military conscripts had developed the disease after vaccination. He concluded that

> it is from having been placed in the presence of similar facts, in other words seeing that the cholera vaccine was far from assuring immunity for those vaccinated, that the Cambodian population no longer has, for this vaccine, the

> enthusiasm with which they welcomed it in 1926. [¶]I will also add that the vaccine equally slid from their initial confidence by the side effects that often followed the vaccination, [which were] such that the conscription Medical Chiefs asked me whether they should or shouldn't utilize the doses being employed.[84]

In the 1930s, with increased experience with both the smallpox and the cholera vaccine, Khmers began favoring the former over the latter. The indigenous *médecin* in Takeo would note in 1930 that villagers came eagerly for smallpox vaccinations, but for cholera, public opinion was quite different.

> The inhabitants have a great deal of repulsion to being vaccinated, they fear the pain and the reaction. On several cholera vaccination tours we have often had difficulty assembling the inhabitants for the process. The majority of inhabitants always find a way to hide or slip away, despite at times the threat of severe sanctions from the *khum* authorities.[85]

In 1931, the Kompong Speu doctor employed a sort of bribery to get people to accept the cholera vaccine. The villagers would only accede to the cholera vaccine if the doctor insisted on performing the cholera vaccination before the smallpox vaccination, which by 1931 most of the population sought willingly. If he performed the smallpox vaccination first, the vaccinated patients would slip away immediately afterward to avoid the injection for cholera. In the early 1930s, the health services occasionally forced the cholera vaccination upon the population during times of epidemic.[86] By 1937, however, only the frontier provinces of Cambodia were using the cholera vaccine, predominantly on a facultative basis.[87]

Even though the smallpox vaccine was more reliable than the cholera vaccine, it too caused problems for the Cambodian health services. As mentioned earlier, French doctors had introduced the smallpox vaccine in 1890; it was the first vaccine employed in Cambodia. Calmette's vaccine institute increased its effectiveness, but the serum was still not entirely reliable. In 1911, the Saigon Pasteur Institute sent a researcher to Phnom Penh to investigate apparent failures of its smallpox vaccine. This researcher concluded that the complaints were exaggerated. He was only able to locate two definite cases of smallpox that occurred after vaccination, noting that in both cases "none of the facts were verified by a competent witness."[88] The issue of competence, also used by Bernard in the Sadec deaths, was often key to the defense of the Pasteur Institutes. In 1923 the mobile vaccinator would comment on the confusion caused by the shortage of the smallpox vaccine during his tour, forcing last-minute postponements and cancellations of prescheduled meetings. Further, the doctor registered many vaccine failures, "the fact of which has not escaped the locals and has been told to me often during my vaccination tours." He also observed the irony that these vaccines—produced by the Pasteur Institute Saigon—were failing a population in need at a time when the administration "request[ed] donations from [the natives] for the 'Pasteur' lottery."[89] During a severe smallpox epidemic in 1928, the smallpox vaccine would prove not just ineffective but lethal. In one village, eight of fifteen vaccinations led to smallpox death. These deaths and the resulting inquiry to the Pasteur Institutes were again kept highly confidential, in order not to shake public confidence in the vaccine.[90] In 1936, when the vaccine received in Phnom Penh tested 100 percent inactive, the institute claimed that the product must have gone bad during its passage from Saigon to Phnom Penh. The health director discounted this excuse, since the vaccine had been delivered in the cool season and used immediately, while vaccines delivered in previous hot seasons had been uncorrupted.[91] Finding themselves without reliable vaccine for an indefinite period, the health services were forced to suspend all vaccination tours in the countryside for the remainder of the year.[92]

An initial resistance to vaccination was neither unique to Cambodia nor to the colonial state. Throughout the world, compulsory vaccination met intense resistance from the general population.[93] In 1899, nearly a decade before the AM would be established in Indochina, the French Academy of Medicine published a brochure on the status of vaccination in France and its colonies. While noting that the French government needed to expand vaccination in the colonies, the academy also observed the many difficulties with enacting mandatory vaccinations in France.

> In France, we have to combat neither the native elite, nor variolation, nor marabouts, nor Muslim fanaticism, nor thébibs, nor the natural defiance of a conquered nation towards the conquering nation, but we do have to struggle against the inertia of the rural population, their preconceived ideas, their parsimony, and their resistance to any action without an apparent immediate utility.[94]

In Cambodia, the choices made by the indigenous population were not simply "defiance of a conquered nation" nor even the "inertia of the rural population," a superficial description of French peasants that could as well be applied (in all of its inaccuracy) to Khmer peasants. Khmer reaction to smallpox and cholera vaccination was not simple resistance or refusal.

Response depended on the nature of the medication and the method of its implementation. The smallpox vaccination, introduced quite early in the colonial period and often performed in a manner echoing previously existing variolation techniques, enjoyed much greater success. Khmers, in contrast, never truly embraced the cholera vaccination and wholeheartedly shunned it within twenty years of its inception, because of its lesser reliability and its negative side effects. In other words, resistance was often learned against a procedure that had at first been accepted. The story of the introduction of the Haffkine plague serum in Cambodia illustrates this process in a single episode.

## Local Choices

As discussed above, two types of plague sera were used in Indochina during the colonial period. One was developed by Yersin and another by Haffkine, a Ukranian working for the British Raj. In 1908, the inspector general requested that all local health services switch from the Yersin to the Haffkine serum for plague prevention. Although its side effects were more severe, the Haffkine serum provided an immunity of several months, whereas the protective effects of the Yersin serum lasted only for several weeks.[95] Whether due to local inertia within the medical establishment or to unavailability of the Haffkine serum, this change was not implemented in Cambodia until 1910. The Phnom Penh municipal doctor, Henri Devy, anticipating that the "faint-hearted" (*pusillanimes*) Cambodians would protest against changed methods, wanted to test the serum on the Chinese population first.

Phnom Penh at this time was a relatively new city. With French encouragement, King Norodom had abandoned the ancient Cambodian capital of Oudong in 1866 and had relocated fifty kilometers downstream to what was then a small town at the confluence of the Tonle Sap, Mekong, and Bassac rivers. In 1907 Phnom Penh was still small, but growing. During the previous fifty years the colonial government had made a substantial effort to organize the capital city of its protectorate kingdom. In an attempt to overlay some order, the administration had earlier partitioned the city into quarters that were denoted by both ethnicity and number. In 1907 there were four quarters, although in later years more would be added.

The Chinese quarter, the most densely populated area, was the commercial center of Phnom Penh. Contemporary photos and correspondence describe it as a place where occupants lived in often haphazardly built and cramped quarters with livestock living among them, garbage and sewage dumped into the streets, canals where rats and other stray unwanted animals roamed freely, and no zoning laws of any sort. The Cambodian quarter was situated at the southern tip of town. The Royal Palace was to be found there, along with the dwellings of Cambodian urbanites. The term urbanite was misleading, since much of the Cambodian population was newly urban and the Khmer population of this quarter was in some ways a crowding of the village lifestyle on the outskirts of the Chinese quarter. North of the Chinese quarter sat the French quarter, and above it the Catholic village, located on the northern end of town. The Catholic and French quarters were physically separated from the indigenous quarters by a canal that encircled them on all sides (figures 2.4–2.5).

Whether by purposeful French design[96] or by the social conditions resulting from colonial economic policies,[97] the city's various ethnicities were, on paper, segregated into relatively discrete geographical locales, although the actual boundaries blurred considerably. The government had attempted to improve canals, build sewers, standardize building codes, mandate weekly cleanings of the open-air markets, and regulate various other behaviors of subject Asian populations—most of this under the banner of improving public hygiene, and most roundly ignored by city inhabitants. The majority of the city's population was not Khmer. A rough census in 1909 estimated nearly fifty thousand inhabitants, with approximately sixteen thousand Chinese and eleven thousand Vietnamese. The French population still numbered under one thousand.[98]

Much to Devy's relief, the first plague case of 1910 happened in the Chinese quarter on January 27. Devy introduced the new Haffkine serum as a preventative vaccination for neighbors of the afflicted. These first injections were innocuous, with only the slightest reaction. Much encouraged, Devy extended the vaccinations to other Chinese, and eventually to Khmers, registering only slight side effects and no cases of the plague. Devy anticipated the "great fear" from the Khmer population that he had already observed whenever a French doctor appeared with any sort of medical instrument. This year, however, a great many plague cases were occurring around the palace, and thus the elite were more open to French intervention. Surprisingly, princes and ministers volunteered for vaccination. After their successful immunizations they sent their secretaries, employees, and domestic servants. From these successes came others.

In March 1910, twenty of 280 monks at the Pagoda Ollalom died of the plague. Devy, accompanied with Prince Sathavong, a very strong partisan of the vaccine, inoculated all of the surviving monks at the pagoda on March 29. "Like magic," the plague ended in the *wat*, and again the vaccinated suffered very few side effects. Because the Buddhist sangha exercised enormous

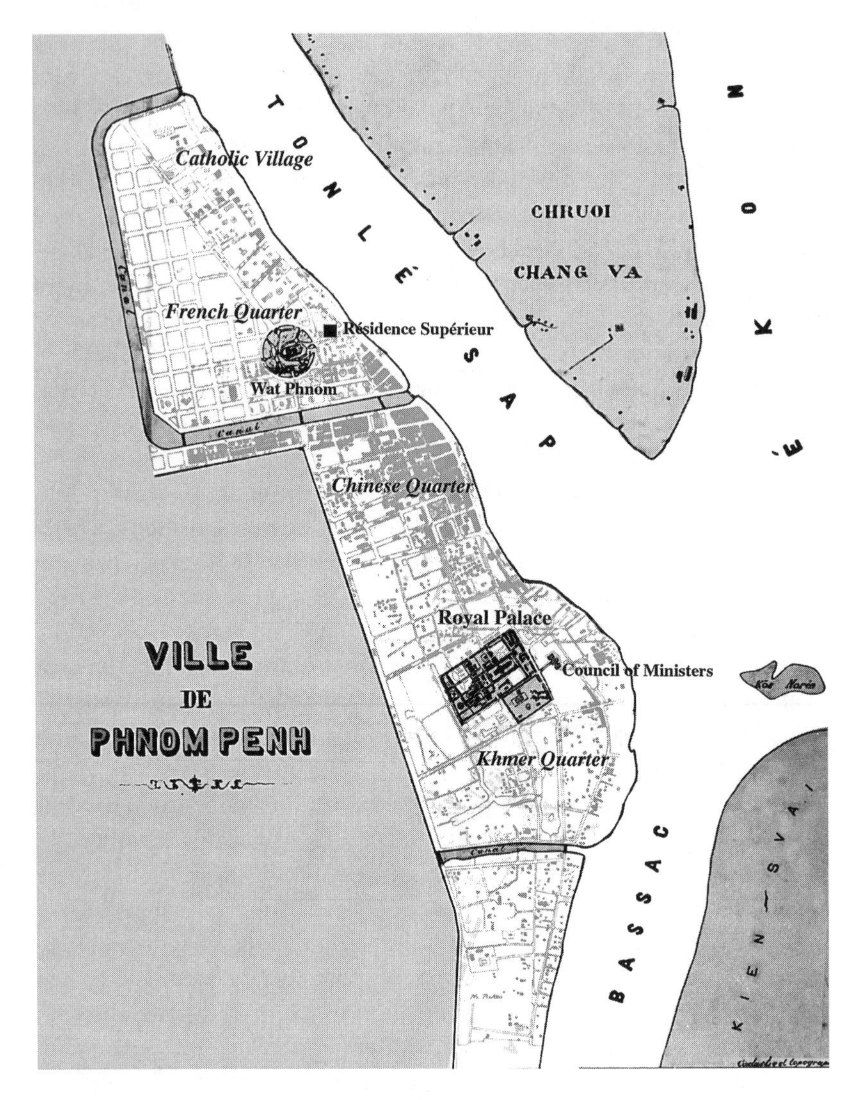


Figure 2.4. The city of Phnom Penh. Author's schematic based on a map produced for epidemic management in the city, circa 1910. NAC RSC 9315. Courtesy of the National Archives of Cambodia.

influence on the local population, the positive word of mouth was widespread. Shortly afterward, the municipal dispensary received crowds of Khmers requesting plague immunization. The serum had nearly run out when on April 20, 1910, the king requested that Devy vaccinate the entire population of the palace grounds. Having only one flacon of serum remaining,

Devy and his staff convened on the palace grounds and vaccinated 517 mandarins, soldiers, and functionaries. The demand was so great that the medical team had to postpone vaccinating the remainder of the staff until the next day. When Devy returned to the palace on the following day, he was met with a large, disgruntled crowd. It seems that everyone inoculated on the previous day had developed high fever, stomach pains, and headache. Devy could do little for them, and though disheartened, he continued vaccinating with what remained of the single flacon of vaccine. However, volunteers were suddenly scarce. On April 21 and 22 he could only find about 150 people who would submit to the Haffkine injection. These vaccinations caused the same negative side effects as of those of the previous days. From that time forward he saw a drastic drop in demand for vaccine. Further, two Cambodians vaccinated on April 21 developed the plague and died a short time later. One of King Sisowath's daughters who had also been vaccinated on the 21st developed the plague but, much to the doctor's relief, survived. Still, Devy pushed forward with the vaccinations. Through the month of May he managed "despite resistance" to perform 1,353 more injections with a new batch of Haffkine vaccine from Saigon. In this month, however, a young woman died a few days after her injection. He dryly noted that her death "succeeded in achieving the ruination of the Haffkine

Figure 2.5. View of Phnom Penh riverfront looking north from the Khmer quarter, circa 1900. The spire in the distance is Wat Phnom, located in the center of the French quarter. Courtesy of the National Archives of Cambodia.

vaccine."[99] Faced with intense resistance from townsfolk, Devy returned to the use of the Yersin serum, noting that this attempted change in the medical arsenal had been "disastrous."[100]

It may appear puzzling why Devy, faced with the failure of the plague vaccine, continued to employ it with such vigor. It would seem that he continued to do this not only because it was the job he had been given, but also because there was little else for him to do about the epidemic but to register deaths. It also bears emphasizing that his actions were neither callous nor confident. Rather, as with many health care workers of the time—and indeed today—he did the work he had been trained to do with little flexibility in response or reflection on its ethical implications. He would be forced to change his behavior not because of intellectual revelations (ethical reasoning or scientific observations), but because of his patients' reactions. Throughout this episode, both the health service and the indigenous populations changed their opinions and behaviors. The indigenous populations did so as a result of the concrete effects of vaccination; French officials did so in response to changes in local behavior.

## Constraints and Privileges

The history of vaccine research in Indochina and its implementation in Cambodia reveals that the concept of "colony as laboratory" must be qualified. The doctors, researchers, and administrators in the French colonial apparatus had considerable privilege over the body of their colonized subjects, but that privilege was tightly constrained. Indigenous response was not always just resistance; it often was a considered reaction to particular experiences. Khmers did not simply refuse colonial medicine; they refused unknown, ineffective, or potentially dangerous treatments. Sometimes they were quite willing to embrace treatments with clear medicinal value, but such treatments were rare in the early twentieth century. Local response, in all of its varieties, influenced and limited French action.

Further, fissures existed among the various scientific factions in the colony. European men operating in the colonies were not, as a group, homogenous in intent or action. While the Pasteur Institute, the AM, and the military medical services were all centralized within the colonial administration, they nonetheless had to maneuver for political priority within the system. Internal and international rivalries and alliances shaped research; ideologies of science could support or undermine ideologies of colonial rule. Further, the dynamics within and between groups—cultural, political, professional, institutional—played an important role. In other words, not only the actors but the relationships between actors, or lack thereof, were crucial. The next two chapters will look more closely at the process and consequences of the expansion of French medicine in the specific milieu of Cambodia. They will examine not only politics within the colonial medical services, but also the operations of colonial medical institutions and ideologies within the social, political, and geographical context of Cambodia.

THREE

# The Politics and Pragmatics of Managing Health

> Indochina was not a French colony, Indonesia was not a Dutch colony, Burma and India did not belong to England[;] the populations of these territories were enslaved by a much more demanding silent tyrant: intestinal worms and parasites of the blood.[1]

As historian Henri Brunschwig observed of the muddle of French imperialistic rhetoric in the late nineteenth century, few people are purposefully hypocritical, and on some level the colonists believed the rhetoric that they used to justify domination.[2] The French imperialist was operating under the dual ideals of nationalism and humanitarianism. Much of this humanitarian ideal was encompassed in the promise of science, and particularly scientific medicine, to improve the lives of those being dominated. Baser goals, such as increased surveillance, greater control, and political domination, certainly informed medical efforts in Indochina. But if the desire of the medical service to improve the health of the indigenous population was only a proximate goal for sectors of policy makers, for at least some practicing medicine it was the ultimate goal. But again, we cannot reduce the often fractious aims within the medical administration too much. Even among the doctors in the field, attitudes varied greatly. The medical chief in the remote conscription of Svay Rieng would write with some frustration, "One only has the right . . . to conquer a country if it follows that one brings to it greater happiness and well-being."[3] In contrast, another medical chief argued in 1912 that the AM's offers of assistance were essentially lost on an unwilling population. His replacement the following year would label such a view much too bleak. French doctors simply had to take a different tack and follow a few guidelines. He urged that "we need to come to them, not as a medical functionary obliged to offer them consultations, but as a man who is interested in their suffering and offers them some relief." If French doctors behaved with a combination of caution and concern, "even the most refractory, the monks and the semi-savage Pnongs, will with great ease permit themselves to be treated [by us]."[4] Attitudes of other doctors, many of whom we have mentioned in passing, ranged from deeply committed to callously indifferent to the well-being of indigenous people.

The Cambodian population, in turn, did not reflexively refuse French medicine. While some Khmers refused it because their needs were already met by traditional practitioners, others made their choices based on their own observations or the experiences of their peers. A doctor refused during an epidemic could be demanded for a broken bone. Certain medicines were accepted and later rejected; others were originally refused and later sought. Many of these choices had little to do with a conscious desire to maintain Khmer traditions or thwart colonial power. The choices of the colonizers were often contingent on the same issues. Rational medical policies created "from above" mutated in the immediate reality of Cambodia. In the end, a survey of the attitudes of French doctors or Khmer patients holds limited explanatory power for the medical situation as it evolved in the country. Intentions, both good and bad, did not dictate results. The encounter between Western medicine and the Cambodian population was driven by many other considerations beyond the will of either party. Human and economic resources, geography, climate, and the scant medical arsenal were a few such factors that affected the evolution of this interaction. This chapter examines the interplay between these factors and efforts to establish the AM in Cambodia.

## Organizational Instability

The scope of the colonial medical services was not shaped only by orders from the metropole or the willingness of the AM doctors. Besides the limited medical options for many diseases early in the colonial period, human and economic resources were tightly constrained. In other words, staff and money were scarce. In the late nineteenth century, the boundaries of modern medical practice were still being delineated. Medical doctors were often also amateur anthropologists, ethnologists, and geographers. For example, French explorer Clovis Thorel served as physician to his fellow cartographer-conquerors while collecting botanical samples with Francis Garnier's Mekong expedition in the 1860s.[5] Yersin mapped the highlands of Vietnam between research trips for the plague in the 1890s. Maurel, who credited himself as the founder of medical services in Cambodia in 1885,[6] became deeply involved in physical anthropology and the short-lived science of craniometry during his stay.[7] By the turn of the century such extraprofessional activities became uncommon, although a few medical researchers would occasionally publish ethnological work on Cambodia. By World War I, doctors of the AM were more discretely bounded within medicine as a field apart from the other life and social sciences. This was due both to the consolidation of medicine as a profession and the stabilization of European life in the colonies. As the colonial apparatus was established and the specially trained European and indigenous population increased, the number of roles any one colonist needed to fulfill decreased. However, while the number of Western-trained doctors rose steeply after the AM's creation, the gap between the number of doctors and the perceived need grew as the AM expanded.

At its inception in 1907, the AM organized Cambodia into thirteen medical districts, although many of these districts would have neither doctors, hospitalization facilities, nor in some cases a simple dispensary until the following decade. From 1907 to 1940—the effective lifespan of the AM in Cambodia—the various medical districts contracted and expanded according to the staff available and the budget provided by the metropole. The two medical facilities existing in 1907 had swelled by 1911 to twelve outlying medical ambulances,[8] the Sisowath Dispensary, the Maternity Hospital, the

**Table 3.1. Assistance Médicale medical districts in French colonial Cambodia**

| District (conscription) | Date established | Comments |
|---|---|---|
| Phnom Penh (Mixed Hospital)* | February 1885 | |
| Kampot | March 1903 | |
| Kompong Cham | August 1907 | |
| Svay Rieng | January 1907 | Folded into Prey Veng in 1910 |
| Kompong Chhnang | March 1907 | |
| Kompong Thom | May 1907 | Folded into K Cham during WWI |
| Takeo | October 1907 | |
| Kompong Speu | 1908 | Folded into Kandal during WWI |
| Pursat | April 1908 | Folded into K Chhang in 1910 |
| Kandal | fJanuary 1908 | |
| Kratie | July 1908 | Folded into K Cham during WWI |
| Prey Veng | March 1908 | |
| Phnom Penh Municipality | October 1908 | |
| Battambang | 1909 | Retroceded from Siam |
| Stung Treng | December 1909 | |
| Siem Reap | 1925 | |
| Haut Chhlong | April 1934 | Eliminated in January 1936 |

*The Phnom Penh hospital was neither a district in itself nor included as part of the municipality of Phnom Penh. Its statistics were reported in the annual report for the entire country.

Chinese Hospital,[9] and the Mixed Hospital in Phnom Penh. In the next twenty years, the number of districts varied with budget, personnel, and political events (see table 3.1).[10]

Despite the global economic downturn in the 1930s the number of districts remained relatively stable, with the temporary addition of Post le Rolland, Haut Chhlong, from 1934 to 1936.[11] The infrastructure remained constant during this period because of two trends: the hiring of indigenous doctors at much lower salaries, and the push for "social medicine" coming from the metropole with the ascension of the Popular Front (which will be discussed in the next chapter). However, the budget deficit did force other extreme measures. Cambodia was temporarily eliminated as a separate medical state from January 5, 1933, to June 9, 1934, when all of its bureaucratic operations were folded into the Cochinchina service.[12]

## A Few Good Men

The instability in medical infrastructure was mirrored by instability in personnel assignments. In some instances the restoration of medical posts was illusory, since they remained unmanned. For example, the health service restored two medical posts in Cambodia in 1917, giving the country a total of nine, a number still woefully insufficient for its size. However, the medical report for 1917 parenthetically notes that four of those posts—more than the previous year—were actually *without* doctors.[13] Thus the French medical presence actually decreased while its institutional representation increased. Available doctors were constantly shuffled between districts as the administration tried to spread its resources as widely as possible. Further, AM doctors did not last long in Cambodia. Alongside the stream of repatriations due to ill health or resignations, with few exceptions doctors opted to post out of Cambodia when given the chance. As Inspector General Paul-Louis Simond observed in 1910, the only posts where AM doctors had any authority or ability to perform useful service were certain stations in Africa or Madagascar.[14] Within Indochina itself, French doctors considered Laos and Cambodia frontier societies, and generally less desirable locales than the three states of Vietnam. Physicians who spent more than a couple of years in Cambodia usually oscillated between two or three posts. The fundamentally transient nature of the provincial doctor's appointment encouraged a social distance, and in some instances an indifference, to his charges. It bears emphasizing that the majority of French medical workers were these men who worked not only at a distance from their patients but also at an intellectual disconnect and social distance from their own professional and social metropoles.

In the 1890s, in anticipation of the expansion of the health services to the local population, the GGI began preparations for a school to train native medical staff. The need for such a school was dictated by the limited financial resources and severe personnel shortages in the health services, since "natives" were both more plentiful and less expensive to train and hire than their European counterparts. The Hanoi École de Médecine, created in 1902 specifically for this purpose, began producing medical staff for Cambodia by 1910.[15] The graduates of the École were not *docteurs*. Having received a *diplôme d'université*, they were only permitted to call themselves *médecins*. Only a *diplôme d'état* (from one of the national medical universities in France) conferred the rank of *docteur*, and the French medical establishment jealously guarded the right to use the term.[16] Initially the École's graduates entered into a separate cadre of AM *médecins auxilaires*. In 1927, the AM created an intermediary category between those of the French and auxiliary cadres, of *médecins indochinois*.[17] The institutional framing of a professional hierarchy simultaneously with a racial hierarchy reflected the racially inscribed assumptions about health care provisioning. As will be seen further, discrete professional and patient categories defined through race would also affect the categorization of disease and illness. While germ theory posited a standard "type" of germ that universally caused a specific disease, the colonial doctor would expend considerable effort describing how the Chinese, Vietnamese, or Cambodian race differentially reacted to and expressed these types. The universalizing aspirations of science fractured under colonial conditions. Thus, diseases became inseparable from their racial expression. The relationship between doctor and patient and the relationship between the germ and its host would be determined by race.

The five separate states of the Indochina federation were expected to provide their own candidates for the École and fund their education through local budgets. In return, these students had to perform a minimum of six years' salaried service for the AM after graduation, usually in the sponsoring state. In any one year the École produced only two or three graduates for the Cambodian service, a number that would actually decrease in latter years. Initially, the majority of applicants from Cambodia were Khmer, but by the 1920s most were ethnic Vietnamese. The RSC received fewer than twenty applicants in any year, and rejected the majority of the applicants as insufficiently educated. This was despite the special consideration shown to most candidates from Cambodia. As late as 1915, all Cambodian candidates matriculated at the École without the requisite *diplôme des études complementaires.*[18] While the number of Khmer graduates at the beginning was already low, by the mid-1920s the number of indigenous *médecins* graduating from the École for all of Indochina began to decrease—a trend that would become worse in the 1930s.[19] In part this was due to more stringent demands on student qualifications; special educational considerations became rare. The few graduates designated for the Cambodian service in this decade were almost all ethnic Vietnamese. The year 1930 would be the last in the French colonial era during which a Khmer candidate submitted an application to the École de Médecine.[20] In an attempt to pinpoint the reasons for this paucity of applicants, the health services surveyed existing Khmer *médecins* in 1932 on their view of recruitment. The respondents were largely negative in their responses; all said they would not let their children follow in their footsteps, as the career was too strenuous, respect almost nil, and the pay too low.[21]

Thus, despite administrative hopes, the École never produced a sufficient number of indigenous *médecins* for Cambodia. As with his European personnel, the health director was forced to stretch his indigenous *médecins* across multiple postings. Those with good performance reviews followed a rotating posting pattern similar to that of their French counterparts, while troublesome employees were frenetically shuffled around the country. The severity of punishment for a negligent European doctor differed from that for an indigenous *médecin*; both were reposted, but the indigenous *médecin* usually also received a demotion and decrease in salary. Indigenous *médecin* Hell, who between 1912 and 1920 was in a series of disputes with *infirmiers* and patients, found himself reposted after each controversy. After a particularly serious disciplinary scandal, the health director sent him to the hinterlands of Prey Veng not in the capacity of a doctor, but as an assistant on a "sanitary mission."[22] One of the most egregious examples of medical malpractice occurred with indigenous *médecin* Kraucht Ketmoung, whose actions against his own countrymen were scandalous enough to make a French newspaper.[23] However, the AM did not fire him, and he continued to be involved in several medical conflicts in the next five years.[24] Ketmoung would complain to his employer,

> Over more than 19 years, I have served at more than 15 medical posts. Since 1934, I have worked at Phnom-Penh 10 months and 2 days, on the worksite of Colonial Road no. 12 for 6 months and 7 days, at Samrong (military post of Siem Reap) 1 year 2 months and 15 days, at Phnom-Penh 2 years 15 days, and again currently at Siemreap with a mobile medical team 1 month 15 days,

> at the Siem Reap hospital 2 months and 14 days, and most recently 20 days in Kompong-Thom. [¶] I have also worked 2/3 of my life in remote and sometimes unhealthy regions.[25]

His plea fell upon deaf ears. Since their contracts stipulated that they had to repay all living expenses and scholarships fronted by the colonial government should they fail to work for six continuous years, indigenous *médecins* were often financially trapped into a commitment. Further, a medical education represented a considerable investment of time: three years of medical school along with six years of AM employment. In turn, the administrative and financial investment the French government made in the recruitment and the training of these indigenous *médecins* and the continual shortage of viable candidates made the administration hesitant to fire them.

Though *infirmiers* were plentiful, administrators almost unanimously viewed them as badly trained, unorganized, and useless in an unsupervised capacity.[26] The job of an *infirmier*, which paid only slightly more than that of a hospital coolie and often entailed the most unpleasant and menial tasks, was only sought out by candidates with the fewest options. Initially, an *infirmier* had to undergo two years of training at a school attached to the Mixed Hospital before entering the AM. The negative assessment of *infirmiers* did not improve during the 1920s and 1930s, despite several efforts at reorganizing the *infirmier* cadres and raising standards of training.[27] To increase the job's prestige, the health services also recruited several royal princes as *infirmiers*. One prince, Cheatavong, was accepted in 1916 but apparently did not complete the program. Of four other princes who applied (Ketaya, Khanthayari, and Rattavong in 1913; Ritharasi in 1918), the first three were deemed insufficiently educated and their applications were rejected. Ritharasi reapplied the following year and was accepted. He only completed one year at the École before making his way to Paris and obtaining a medical *diplôme d'état* from the Faculté de Médecine de Paris in 1929. Unfortunately, his status as a Western medical doctor did not seem to raise the social status of either *médecins* or *infirmiers* in the AM. Rather, his return to Cambodia would be a thorn in the side of the medical administration. The AM was pressured to hire him briefly as a contractual doctor but did not renew the contract. Ritharasi would publish several small medical trivia pieces in the *Echo du Cambodge* in the 1930s. His medical thesis on traditional medical practices in the country is one of the earliest descriptive works of precolonial Khmer medicine.[28]

The health director had made similar efforts in recruiting royalty to undertake training as indigenous *médecins* in Hanoi, though no prince ever graduated from the École de Médecine.[29] The Khmer crown perhaps too fully cooperated with these efforts, for by 1927 the health director would complain of the incompetent princes he had been forced to hire as *infirmiers* due to pressure from the royal family to aid these "idle" men.[30] Colonial administrators were attempting to borrow the sacred prestige conferred on the charismatic healer or royalty and associate it with a new bureaucratic, secular elite. In medicine more than in other newly created civil services, bureaucratic or technocratic authority alone did not enable the workers to do their jobs. Ministering health was fundamentally different from administering records. The power to heal had been tied to personal charisma, merit, or karma. Such traits were not necessary to get through the few French-language schools of Phnom Penh, the new routes to opportunity and power created by the colonial government. Thus, respect for the role of a Western doctor never materialized among the indigenous population,[31] and salaries remained low while training became more strenuous. Efforts to tie medicine to royal prestige ultimately did not make the job of *infirmier* or *médecin* more desirable.

## Racial Discord

The shortage of personnel was made more problematic by the ethnic mix in Cambodia—or, more accurately, by the ethnic categories constructed by the French. These racial categories, their reification in administrative and bureaucratic structures, and the differential treatment of these various groups heightened Khmer dislike and resentment of the French medical system. Within the AM personnel structure, doctors and *infirmiers* were divided into separate European and indigenous cadres, with a considerably higher pay scale in the former. "European" and "indigenous" were not unproblematic categories. For example, Pondichery-trained[32] Krishim Ittiacandy had an anomalous career in Cambodia that revealed the instability of racial definitions in the AM cadre. At least partially, if not entirely, Indian by birth, Ittiacandy was with the AM through its entire life span in Cambodia. He entered the service in 1907 as an *infirmier* in the French cadre. Having received no additional formal training, he was nonetheless by 1913 serving a role equivalent to that of an indigenous *médecin*, as the medical chief of Takeo district. When posted outside of Phnom Penh, he served in the role of medical chief in the indigenous cadre. When he served at the Phnom Penh Mixed Hospital with French doctors, he was referred to as chief *infirmier* of the French cadre.[33] His race and his professional status were not determined by inherent biology, appearance, or technical skills, but by the number of individuals present within a given racial-professional rank. In other words, as an Indian among Indochinese, he was charged with the duties of a doctor. As an Indian among French medical staff, his presumed capabilities and duties diminished considerably. With the advent of Vichy, Ittiacandy quit the AM and requested a license to practice as a civilian doctor. Despite his lack of diploma, the Vichy government in Indochina granted his request, reasoning that the number of Indians in Phnom Penh was sufficient to warrant an Indian doctor.[34] Again, the colonial government assumed that as an Indian, Ittiacandy would serve Indians. The assumption that "like treats like" underpinned and ultimately undermined French efforts at medical recruitment. As we will see, other aspects of social medicine and women's health services were affected by similar ethnic and racial tensions.

Although resentment between French and the "other" existed, particularly for the *métis* (mixed) population, the greatest friction in the service arose between Vietnamese and Khmer personnel. Less than harmonious as peers, their relationship as patients and practitioners was even more strained. Racial tension only increased through the 1930s. In his 1932 report the health director ascribed the indigenous population's rejection of French medicine to racial animosities. The inspector general, in a blistering response, censured the health director for his excuses, informing him that the role of French medicine was sanitation, public health, and the control of epidemics—not racial politics. He wrote, "Our role . . . will . . . not be hindered or sterilized by the racial incompatibility asserted by all of those with experience in Cambodia and the results obtained there up to now."[35] The following year, the health director requested that the inspector general lower the educational standards of the École de Médecine in specific circumstances to admit more Khmers, who were at a disadvantage in terms of primary- and secondary-school education. He urged that this change and the resultant increase in Khmer doctors would be the only way the AM could get the local population to move away from its traditional healers and embrace French medicine.[36] The inspector general again rebuked the health director, arguing that such ethnic favoritism was unfair to more qualified Annamites. The problem was not simply in the medical service, for the Khmer elite were becoming more vocal in their complaints about the increasing number of Vietnamese in all branches of the Cambodian government. The RSC, echoing the health director on many points, wrote to the governor general arguing that

> it would be easy for me to respond that until now the Cambodians have not made the effort necessary to successfully earn the diplomas that would permit

> them to obtain the jobs they desire, but I perceive clearly that national pride suffers from this situation and that the Protectorate is not far from being accused of insufficiently assisting the reigning party.[37]

The problem was essentially circular. The Vietnamese, perceived by the French as more open to education and quicker learners, had received the bulk of the colonial government's efforts at improved primary- and secondary-school education in the late nineteenth century. French efforts to create and improve the educational system in Cambodia were a relatively late development. The French opened the first *lycée* (high school) in Cambodia at the late date of 1936.[38] Further, the École de Médecine, the only school to train indigenous AM *médecins*, was located in Hanoi. Although the instruction was in French, Khmer students also had to learn Vietnamese, live in dormitories with Vietnamese students, and immerse themselves in Vietnamese culture for several years. This served as a further strong deterrent to Khmer interest in Western medical education.[39] The AM, like other branches of the French colonial government, found it more convenient and cost-effective to funnel educated Vietnamese into necessary positions. In 1938, Cambodia's health director informed several provincial *résidents* that, despite their numerous requests, ethnic Khmer *médecins* and *infirmiers* could not be provided because the administration simply did not have enough supply to fulfill the demand.[40] Of thirty-three indigenous *médecins* employed in Cambodia in 1938, only thirteen were Khmer.[41]

Of course, real differences did exist between Khmer and Vietnamese culture. Further, the animosity between the two groups predated French colonization.[42] However, the French tendency to reify cultural characteristics into bounded ethnic "species" and increased Vietnamese migration into the country, encouraged by French economic and social policies, amplified the differences and increased Khmer resentment. However, French racism was of an inconsistent sort. While scientists carefully tallied characteristics of individuals—head shape, fecundity, intelligence, personality, strength, and so on—and fecklessly turned them into population traits of discrete races and tribes, they showed little interest in the less quantifiable if more significant differences in culture. For example, the government spilled much ink on the regulation of Sino-Annamite medicine in Vietnam and Cambodia, but made no distinction of Khmer, Lao, Cham, Hmong, or other pharmaceuticals.[43] Inspector General Lucien Gaide justified a 1925 study of Sino-Annamite pharmacopoeia in Indochina by arguing, "The understanding of plants and other substances used by indigenous healers is in effect intimately bound up with the understanding of the country itself, of its flora, its language and its customs."[44] Yet the study's scope excluded all Asiatic products not "sino-annamite." The RSC observed that in Cambodia this was an odd omission, since that year the colonial government's Mixed Hospital had sold 128 Khmer or Sino-Khmer pharmaceutical products that did not fall into the category of "Sino-Annamite."[45] The search for culture solely through Sino-Annamite medicine prioritized Chinese and Vietnamese products as "Indochinese." The omission of the Khmer medicines of Phnom Penh's Mixed Hospital reflected the tendency of the Saigon-based colonial administration to conflate "Annamite" with "Indochinese." As the concept of "Indochinese" was at this moment still being constructed, this omission also reflected the process of erasing ethnic diversity.

## Miles to Go

The sparseness both in medical establishments and in the personnel to man these posts was compounded by the region's physical environment. The reality of Cambodian geography made distances even more distant. Before the colonial government had established many of its new roads through the country,[46] Cambodia had no real transportation infrastructure, and distances of a few kilometers could be daunting journeys over flood plains, jungle, and rivers. Geography served as a physical barrier to the commingling of Khmer and French in much of the countryside—a barrier that both parties had to exert considerable effort to overcome. The mobile vaccinator in 1907 would describe days of marching through the brush in the tropical heat without encountering habitation, and then finding, to his frustration, isolated villagers fleeing at his approach. One doctor described his efforts to reach a small village that had requested his services: "It was necessary to cross a forested region on trails that were difficult to see and follow, blocked by torrential streams that could only be crossed with bamboo rafts made on the spot, or sometimes swum."[47] The hazards were not understated. One doctor drowned while attempting to cross the Mekong during a regular vaccination tour in Kratie.[48] A villager who desired vaccination could go to the Frenchman, taking himself to an ambulance or a preannounced *séance locale*, but peasants too had to travel for days to reach such a center.[49] Rural inhabitants not only lived too far from existing medical posts, but the sick also could not afford transportation, usually in the form of an oxcart or boat ride, to the nearest doctor. The district medical doctor, who was often the only doctor at a medical post, could not cover much ground in his monthly trips into the hinterlands.[50] The medical chief of Kompong Thom, one of the largest and most sparsely occupied districts, observed in 1910 that

> the biggest obstacle to the progress of the Assistance is the distance that needs to be covered. One can hardly imagine that a gravely ill patient would undertake a journey along difficult roads, traveling 3, 4, or 8 days by the only means of travel available during the dry season, in a cart without suspension, pulled by horned animals; even those with only minor ailments think twice before undertaking the trip.[51]

When the number of Kompong Thom ambulance consultations doubled from 1925 to 1926, the medical chief ascribed the increase to the recent improvement in the colonial roads, for the natives "would prefer to suffer among their own than to face the difficulties of the road."[52]

By the 1930s, Cambodia had obtained at least some rudimentary routes of travel. However, transportation problems still factored into the lack of interaction between Western medicine and the indigenous population. In

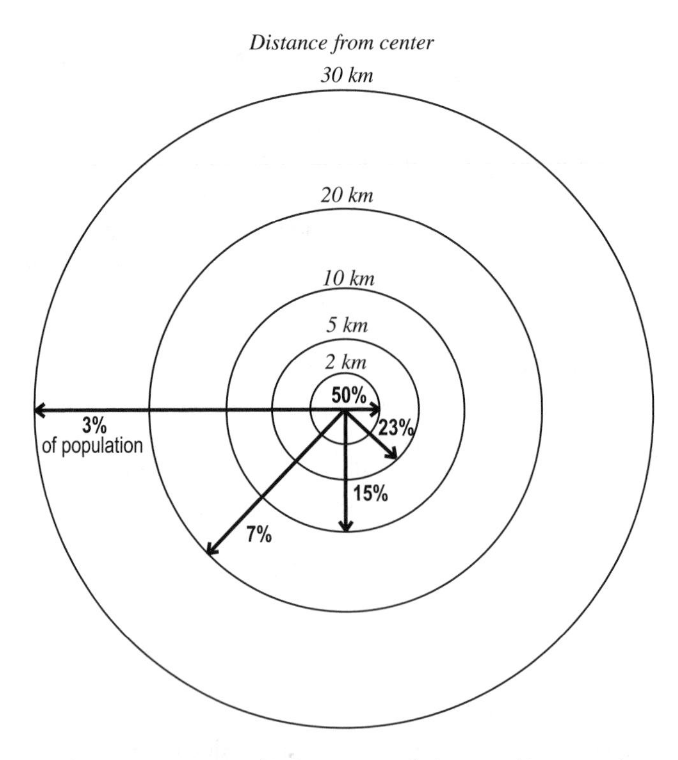


Figure 3.1. Percentage of indigenous population consulting a French doctor in relation to distance from medical center, 1938 (not to scale). Author's schematic of data from file NAC RSC 605.

1938 the health director undertook a detailed examination on the relationship between routes of transportation and the influence of the AM on the population. Taking data from a study done in Kompong Cham by the short-lived hygiene sector and a similar study done in Annam, the health services found that within a distance of two kilometers of the medical center almost 50 percent of the indigenous population had consulted a Western-trained doctor, whereas within a distance of thirty kilometers, only an estimated 3 percent of the population had ever had recourse to the AM (figure 3.1).[53]

## Medical Choices

The limited popularity of French medical care was not only due to distance. The Khmer villager preferentially self-treated even when Western medicine was readily available. For example, one evening in 1925, a Holman snake bit a young Cambodian man—not an uncommon occurrence in the Khmer countryside.[54] This young man had a cousin who worked for the French militia. The Battambang doctor who would eventually see the victim noted with considerable sarcasm that having been bitten by the snake at around eight in the evening, "naturally, he didn't hit upon the idea of going to the hospital, he returned to his cousin, adjutant to the militia, to request that he treat him. This adjutant declined to do so, but he brought us the victim around 21 hours and 30 minutes."[55] In Kompong Chhnang the attending doctor complained, "How many sick come for consultation only when they have reached a lamentable state, after having exhausted all the tricks of Asiatic pharmacopoeia and sorcery . . . . Then, after each 'black series' of 2 or 3 deaths, one senses a certain hesitation among the sick."[56] In other words, Khmers went to Western doctors as the final resort; thus, these doctors saw the most severe cases and had the lowest perceived success rate. In 1931, Dr. Yinn Vann in Takeo echoed this sentiment: "The cases that we are presented with most often are advanced tuberculosis with cachexia. The patient comes to us after having tried all sorts of native empirical medicines and when they despair of being healed, they decide to enter the hospital."[57] In Kampot, the medical chief noticed that indigenous functionaries of the French administration, *including the hospital staff,* would openly request time off to get medical treatment outside of the hospital, avoiding the use of Western medicine.[58]

Again and again, French administrators mention the timidity of the Khmer population vis-à-vis Westerners, and particularly Western doctors. Certainly, the indigenous hesitation was not unfounded, as some colonial doctors did not care to be bothered by native clientele. Many AM doctors seemed to find their patients a source of irritation. In 1909, the *résident* of Prey Veng complained to the RSC about the indifference of local medical chief Paul Bonnigal to his duties to indigenous populations.[59] Bonnigal argued that he was available to those who approached him, and that it was not his place or responsibility to force himself on native life. Pujol

> asserted to doctor Bonnigal that some of his colleagues, whose names are well known, profess a great deal more regard for the natives than he did, that they include in their daily rounds the Cambodian villages in which their Residence is situated. . . . [Bonnigal] seemed to dismiss the moral obligation of a doctor in the public health service to give aid whenever it is needed, to the sick who may be ignorant, afraid, and who dare not ask for our assistance because they fear approaching us.[60]

In contrast, other AM doctors wanted to expand the sphere of Western medicine. Dr. Nicod urged the placement of a dispensary outside of the military camp of Sisiphon so that the villagers would be less afraid to approach them for free medications.[61] Dr. Bazile-Joseph Dupont, in his rounds in the brush, noted the superstitious nature and timidity of the rural people, claiming that the bad treatment of one native very quickly led to avoidance by all:

> In one village shortly before my arrival, a European had brusquely reprimanded his Annamite boy, but without hitting him. Coming after, I could not but feel the main repercussions of the negative effect it produced on the spirit of the Cambodians; all the inhabitants fled and didn't return until two days later; questioned, it was not difficult to get them to confess that they were afraid because of the other European. [¶] These facts hardly make an impression on other civil servants; the doctor alone notices it, because he alone is uniquely made to call upon the voluntary will and the sympathy of these populations.[62]

The medical staff recognized that health care provisioning, unlike many other colonial services, required some voluntarism on the part of the patient. In 1912, Dr. Louis-Henri-Marie Galinier noted that in his rounds in the brush of Kompong Speu, he almost always had to go to the sick patients, as they were much too timid to approach him during his open consultation hours in the villages. With some resignation he noted that "the native is afraid of the European, and trusts in his own peers. Numerous are the sick who wait for the departure of the doctor before they treat themselves."[63] However, the "European" status of the doctor was not the only reason for aversion. In this same year, the steep mortality rate at the Chinese hospital—which was attended by Chinese doctors—was blamed on the condition of the entering patients, who would only resort to it as "a last extreme."[64] Further, the addition of indigenous doctors to the AM did little to attract indigenous clientele. As many of the above examples reveal, indigenous *médecins* often received the same cold shoulder as their European counterparts.

## *Médecins* and Medicines

Considering the formidable nature of travel, the sick frequently chose not to come to district centers, where most medical establishments were situated. Often they sent family or friends on the arduous journey to request medicines on their behalf, much to the irritation of the French doctors.

> It is absolutely exceptional when a local notable, and especially one of our students, secretaries in diverse administrations, nurses, etc. appeals to the doctor; however, all have recourse to various pharmacopoeia that they ingest without the slightest concern for either . . . diagnosis or possible special circumstances in their case. . . . There is thus no need to be astonished if consultations are often given for [medical] "procuration" to a friend or parent delegated by the patient.[65]

In the 1927 annual report, the medical chief of Kampot observed, "If medicine generally has had little success, European pharmacopoeia on the other hand . . . has had enormous success." The Phnom Penh doctor would second this observation, noting with some annoyance that "the mentality of our native patients is still quite childish: it is not advice that they come to demand, it is medicine. It is not the doctor they come to see, it is the 'provider of medicine.'"[66] Throughout the colonial period, medical chiefs would observe, in a single report, the unpopularity of French medicine and, shortly after, the overwhelming demand for French *medicines*.[67]

While the difficulty in travel for a sick individual no doubt encouraged procurement by secondary parties, a second and equally powerful reason for the Khmer view that the French doctor was simply the "provider of medicines" likely involved the centrality of ritual in many indigenous healing practices. As anthropological studies of material culture have emphasized, objects have varying abilities to move among different cultural contexts (exchangeability). As one study observes, "Objects are not what they are made to be, but what they become."[68] The empirical herbalist, perhaps the closest equivalent to a pharmacist in modern medical nomenclature, was the only person in the Khmer pantheon of healers who often gave medicines without performing at least some sort of ritualistic action. This action could be a ceremony involving an entire village over several days, or simply spitting on an afflicted part with a few significant words. The absence of such ritually significant action, or its replacement with a few questions followed by wordless palpations—acts of diagnosis separate and distinct from the act of healing itself—would place the French doctor very much into a position analogous to that of the herbalist, albeit a powerful herbalist, in Khmer medical hierarchy. Divorced of rituals, pharmaceuticals were artifacts, and artifacts could be separated from their original purpose and imbued with an entirely different set of significations. In other words, the social meaning of the pharmaceutical was more malleable than the social meaning of the doctor. As one study of modern medicine observed, because medicines can be transported so easily, they are often "recast in another knowledge system" that differs from the regime of "expertise that developed, produced, and prescribed them."[69] While historians tend to emphasize textual and locutory expressions as central to understanding historical processes, here is an example of the silent but significant role of ritual action—what Bryan Pfaffenberger terms "nonverbal communication as coordinator of technical activities."[70]

The demand for Western pharmaceuticals was manifested in other ways. On many occasions, villagers of their own accord would collect donations and build local dispensaries, after which they would request personnel and necessary equipment from the health director. In some instances, the health director was willing to reassign an *infirmier* from the closest medical post.[71] If no staff was available, these dispensaries lay idle, and occasionally the health director would reallocate equipment to more populated centers.[72] In an exceptional case, townsfolk in a small village of Kompong Chhnang province built a dispensary on their own, but were refused any personnel or equipment because of the locale's remoteness and the lack of available staff. The villagers then found a young man from their own small village to travel to Phnom Penh to complete the two years of *infirmier* training, an act impressive enough that the RSC allocated four hundred piastres to equip the village dispensary after the man's graduation.[73] Thus, in certain instances indigenous populations not only embraced Western medicine, but were willing to expend considerable effort to obtain new technologies with perceived benefits. However, the health services were skeptical of these posts.

> The extension "on the surface" of services of the Assistance, in other words the creation of numerous small medical posts confided to infirmiers placed under an often illusory supervision, is a bad thing, and also too frequently in certain cases, a real danger to the population. . . . The few badly digested and always misunderstood notions of pathology and therapeutics that form the professional knowledge of the indigenous infirmiers is sufficient to give them a [false] assurance and self-confidence that the sick too often pay the price for.[74]

These scattered posts also meant an additional surveillance trip for the nearest medical chief, and again the difficulty in transportation and the shortage of secondary staff created a problem.[75]

## Necessary Compromises

The attempt to train Khmers to appeal to Khmer patients was ultimately driven aground by the lack of appeal of the medical profession for the average Khmer man. Years of arduous medical training in a foreign country led to a salary roughly equivalent of that of an administrative secretary in the French bureaucratic apparatus, a position that in contrast required no post-*lycée* education. As patients, Khmers were not simply refusing "new medicine"; they were also rejecting an arduous road trip for doubtful treatment by individuals with whom they could rarely communicate. They were also resisting procedures that made little cultural sense to them. The local health services quickly realized that other techniques would be required to appeal to indigenous populations and increase the scope of the medical services within the AM's operational limitations. Since the Khmer population did not prove as open to French medicine as originally envisioned, the colonial service had to display some flexibility.

French attempts to accommodate the culture within which it was operating can be divided into three categories: limiting local competition through the elimination or regulation of indigenous healers, incorporating indigenous customs and medicines within French medical practices and establishments, and popularizing French medicine at the grassroots level with education and propaganda. We will finish this chapter by examining the first and second categories. The third category will be considered in the next chapter in conjunction with colonial attempts at "social medicine" and efforts towards public health, which were also referred to as "sanitary defense."

## Eliminating the Competition

In 1921 the health services, in one of their sporadic efforts at pharmaceutical regulation in Cambodia, requested that the civil administration in each district provide them with a list of native healers. Most districts returned short lists of native medical practitioners, and two districts (Kompong Cham and Pursat) claimed to have none. Considering the daily experiences that the medical staff had in competing with local healers, these lists without a doubt grossly underrepresented their vast number. It seems the civil administration was either ambivalent or indifferent to the medical administration's desire to compile the lists.[76] The health director, in a letter to the RSC, tried to allay fears in these outlying districts.

> The individuals who have declared themselves doctors may continue to practice their profession as before. They will disappear by extinction. Monitoring will take place to this effect in each Residence. But, in the future, no one may exercise empirical medicine if he has not obtained the authorization in a form that your decree will fix. . . . We do not want to prevent the natives from resorting to healers, bonesetters, or others from day to day. We simply desire the ability to exercise a bit of surveillance over these individuals in order to prevent them from harming their peers.[77]

Ultimately, the health director's vision for the future was too ambitious. Throughout the French colonial period, native medicine showed no signs of any tendency towards "extinction." We have no record of the RSC limiting the practice of empirical medicine, although there was much legislation on the regulation of Western medicine and its practitioners.

Considering that the number of AM doctors began in 1907 at four, and by 1939 had only reached some twenty-nine Indochinese *médecins* and nine French doctors,[78] eliminating local healers would have been a practical impossibility. The estimated population of Cambodia in 1939 was slightly over three million.[79] Without the presence of practicing traditional healers, the ratio of doctor to patient would have been one to seventy-nine thousand. The attempt to cultivate civilian doctors (outside the AM) was also a failure, simply due to the economic state of the country. By 1939, only six former École de Médecine graduates, all ethnic Vietnamese, were operating private medical practices in Cambodia.[80] The AM, always strapped for cash, was itself unable to recoup even marginal costs from the indigenous populations. For instance, during World War I the French administration began to investigate patients at the Mixed Hospital to confirm that they were indeed indigent before allowing them free medical care. The municipality also hoped to recoup some fees from the patient's hometown if the patient himself was unable to pay. Two policemen manned the hospital's consulting room, interrogating incoming patients as to their status, job, and town of origin. The health services collected little scraps of information on thousands of patients, most confirming that they were indeed penniless.[81] Ineffective as a means of collecting money, the investigations were discontinued in 1920. However, the global depression forced the AM to reinstitute the measure in 1933.[82] This practice was highly unpopular, making the hospital experience not only frightening but also embarrassing for the patient. Further, if the administration attempted to collect the fees from his local village, the patient often faced recrimination by his village chief.[83] At least two private newspapers recorded the intense patient dislike of these measures, and mounted a campaign to eliminate the practice.[84] Nonetheless, the investigations continued through 1937, at which point RSC Léon-Emmanuel Thibaudeau again rescinded them.[85] Pennies could not be squeezed out of the penniless.

In 1940, most of the population were still subsistence farmers. They could not participate in the contractual cash economy demanded by modern medicine. In contrast to AM doctors, most traditional medical practitioners did not rely solely on their medical talents for survival. They themselves were also farmers, monks, housewives, and so forth. For them, medicine was not so much a profession as one of a variety of social roles they had to fulfill within their villages. The colonial doctor and the indigenous healer were operating in distinct economic and social systems. While the Khmer patient did not have the financial capital to participate in the French medical system, the French doctor did not have the social capital to integrate into the Khmer system.

## Medical Pluralism

Legislation and strategic plans produced by the GGI and the AM reveal few instances of organizational interest in the specific conditions of Cambodia before the 1930s. However, if we examine the personal communications and daily records of administrators in Cambodia, we do find that some showed remarkable sensitivity to the indigenous worldview on medicine. Certain administrators attempted a combination of medical pluralism, outreach, and accommodation. A paragon of this effort was an administrator by the name of François-Marius Baudoin, *résident* of Kompong Cham 1905–07 and later RSC in 1915–18 and 1924–27. In 1905 Baudoin, with the aid of his medical personnel, compiled a practical medical guide that he then distributed to area village chiefs.[86] Completely voluntary and dispensed with the added incentive of free medications, these guides were very popular, reprinted several times over the next few years. *Résidents* in neighboring provinces also distributed them.[87] The guide itself was a twenty-page eclectic mix of practical information, ranging from methods of using French pharmaceuticals such as boric acid, potassium permanganate, and quinine (all transliterated in the Khmer alphabet), to means of preventing cholera and reviving drowning victims.[88] Kompong Cham was one of the most successful medical districts, if we measure success by the percentage of population that both embraced and received some benefit by French medicine. Due to the lack of true centralized control within the AM, as well as to the limited budget within the medical services, the civil administration's interest—or lack thereof—often determined the reach of each medical district chief. Baudoin, for instance, paid for these brochures out of the civil administration budget. Further, he himself approached the medical staff as well as the RSC to have free medicines distributed with these brochures. His early interest in the health of the Kompong Cham population was reinforced by the arrival of Bernard Menaut, a dedicated physician who would spend his entire career in Cambodia, most of it in Kompong Cham.

Baudoin's acceptance of various Khmer medicines would cause some friction with the AM bureaucracy, again illustrating conflicting priorities between the layperson and the medical professional. His local medical staff may have cooperated with his efforts, but they did not consult the Cambodian health director beforehand. In 1907 Baudoin, furious about the recent directives of the Cambodian health director to his provincial medical chief, sent a confidential letter to RSC Louis-Paul Luce. The health director had ordered the Kompong Cham doctor, Pierre-Georges Lannelongue, to replace Baudoin's medication for cholera, the "Khmer Mixture," with a French medicine called the "Chastang Potion."[89] As the name indicated, the Khmer Mixture was a preexisting native remedy for cholera: a mix of water, alcohol, and other likely inert substances. It had been distributed over the past three years and had proven relatively effective. Baudoin fumed that Hauer's reason for replacing the medication—because the mixture "was not formulated by a doctor" as required by regulations—was asinine, since the health services were ignorant of both the composition and the inventors of many French medicines circulating in Cambodia at the time. Truly, it would seem that a certain medical chauvinism was at work; the likely effectiveness of both Potion and Mixture dwelt only in their ability to rehydrate the patient.[90] Neither acted on the cholera vibrio itself. And yet the

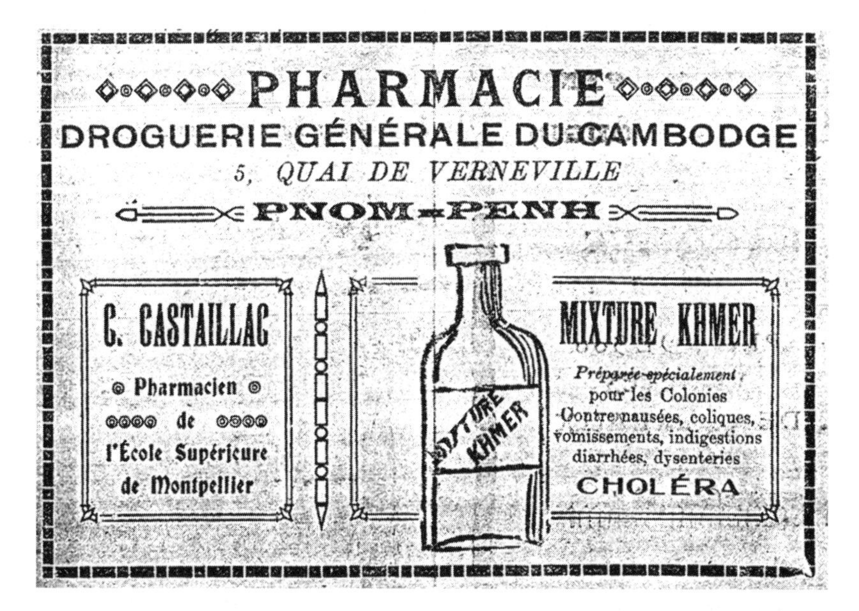


Figure 3.2. Advertisement for the Khmer Mixture. *Echo du Cambodge*, French edition, 1925. The mixture was never advertised in the Khmer edition. The advertisement stopped running in 1927. Courtesy of the National Archives of Cambodia newspaper collections.

Khmer Mixture was significantly cheaper than the Chastang Potion.[91] Baudoin temporarily won his point and was allowed to continue distributing the Mixture in Kompong Cham. However, the Potion became the standard cholera medicine used by the AM after 1907, and it would be the only medicine officially used by the AM for cholera until the arrival of the largely unsuccessful cholera vaccine in 1919.[92] A few years after Baudoin's departure from Kompong Cham, medical chief Menaut would revise his brochure to replace instructions on use of the Khmer Mixture with instructions for the Chastang Potion.[93] Administrative distribution of the Mixture also ceased at this point. However, it seems to have remained relatively popular with the general public, including the French population. As late as 1927, a French pharmacy openly advertised the Mixture to the French-speaking public in the *Echo du Cambodge* (figure 3.2).[94]

Baudoin's efforts were laudable but, in the end, not indicative of any general stance of the French government towards its colonial subjects. Flexibility and accommodation were not guiding principles of the AM, but were often a localized reaction to circumstances in which neither force nor bribery was effective. The administration could not simply beat or coerce health into a patient. Better health, in the end, had to involve some voluntarism on the part of the client. A short history of the hospital, the site of French health care, reveals the different lines of pressure pushing its transformation from its idealized Western form into something just a little more localized.

## Care and "Carence"

In July 1885, the large number of wounded generated during the insurrections that had begun the previous year necessitated the creation of the first extended ambulance in Cambodia.[95] This ambulance would serve as the beginnings of Cambodia's first hospital.[96] In the original months of its existence, the ambulance only admitted Frenchmen; Annamites wounded for the French cause were hospitalized in the Catholic crèche, run by the Sisters of Providence (Soeurs de la Providence).[97] Three months later, the governor of Cochinchina allocated Cambodia the modest sum of four hundred piastres to build a facility to treat natives as well.[98] With additional funding from the local budget and donations from the private Association des Dames de France, the medical staff erected a hospital with separate pavilions for French officers, French soldiers, and natives.[99] Dubbed the Hôpital Français or the Hôpital Militaire in the first two decades of its existence, the building catered to French and Asian soldiers in the French colonial army, and also to French colonists. By the advent of the AM, the establishment began receiving more indigenous patients and became known as the Hôpital Mixte, or Mixed Hospital. The term "mixed" referred to the availability of care to both indigenous and European; however, not much true "mixing" occurred. The health director argued that this was in large part for the benefit of the locals. There was "a pavilion adjoining the hospital, everything in it completely independent, so that the Asiatics may come there without having fear of encountering Europeans."[100]

Fear of encountering Europeans in the hospital was not the only thing that discouraged Khmer entry. The hospital was not only a site of foreign and colonial authority; it was also the site of the doctor's professional authority. In the West, the hospital had begun in the late eighteenth century to undergo a process of transformation from a disordered site of charity for unfortunate social beings to a site of highly rationalized management of the patient's body. Charles Rosenberg has argued that through the nineteenth century, in the American hospital the patient was transformed from an inmate (as in an almshouse) to a patient (with the increase of acute care and decrease of chronic care) to a disease carrier (with the rise of accurate laboratory diagnostic tools).[101] Faure describes a similar process occurring in France, noting that through the first half of the nineteenth century, the hospital was feared as a site of poverty and death.[102] The hospital as a site for healing rather than dying was something new to late-nineteenth-century Europe. Although the hospital administrators in Cambodia may have desired to be modern-minded and "forward-thinking" to match trends in the metropole, they were often forced to allow the hospital to "devolve" into something that allowed the patient to remain a social being and a family member.

Hospital facilities in Cambodia were at first makeshift endeavors—particularly the provincial hospitals, which were often just dilapidated huts with a few cots, lacking running water and electricity. In 1940, at the tail end of the AM's lifespan in Cambodia, the health director would still note that the only "real" hospital in Cambodia was the Mixed Hospital; the provincial buildings could only be considered medical centers at best.[103] The idealized sterile, regulated, orderly environment of the hospital simply was not possible with the resources available, and in the face of specific problems of local environment the staff had to confront. As late as 1928 even the Mixed Hospital, this "most modern" medical facility in Cambodia, would have a serious problem with monkeys defecating from the building's rafters.[104] Further, the transformation of the "social being" into the compliant disease carrier had to await the transformation of the surrounding population into willing hospital patients.

In 1909, to get natives deeply distrustful of hospitalization to stay in the Kompong Cham hospital, the medical chief decided to waive all hospital regulations. Visitors were allowed into the building at all times, day and night; patients could be fed either by the hospital staff or by their family; and native customs and certain rituals were permitted on the hospital grounds.[105] However, the indigenous fear of the hospital was not simply a fear of the French doctor or of the obstruction of necessary rituals. For instance, the military doctor of the rural area of Sisiphon had a particular Khmer patient with an ulcerous sore covering his entire right leg. This patient had come a considerable distance to see the doctor, and his wound needed extended treatment. To the doctor's exasperation, the man ultimately chose to live under a nearby tree for the next six months, coming in regularly to consult the doctor rather than entering the infirmary, where he would have had a "decent" bed free of charge.[106] Clearly, in this patient's case, daily consultations with the doctor were unproblematic, while confinement was to be avoided at any cost.

Urban Khmers also showed themselves resistant to hospital confinement. In 1910 the attending doctor at the Mixed Hospital would grouse,

> The Cambodian wants at all times and all places to do as he pleases: leave when he wants, return at whatever time he wants, eat what he wants, treat himself as he likes, such things all totally incompatible with the effective running of a health establishment; under such conditions, he prefers to remain among his own, whatever the consequences.[107]

For the same reason, the indigenous population also refused surgery. Once the doctor brought down their temperatures, patients with severe infections departed in haste, often despite the still obvious need for surgery.[108] The patient not only associated the hospital with the possibility of his own death, but also feared it as a site of "the dead." The Siem Reap maternity, built in 1933, witnessed only one birth in its first year of operation. This was in large part because it had been built directly behind the section of a pagoda where the remains of the dead were kept.[109] Giving birth at a site of death was taboo.[110] This may have been compounded with the association between the maternity, the cemetery, dead mothers, and fetuses; the angriest ghosts in Khmer folklore are those of the unborn fetus and the mother who died in childbirth. In provincial districts, many of the medical centers did not have morgues. In 1937, the Kompong Thom doctor would observe that when a body was left in a bed overnight in a common room until the gravediggers came in the morning, the other patients, if conscious, would desert the hospital regardless of their physical condition.[111]

Regulations and discipline involved more than the rationalization of medical authority for the sake of curative efficacy. Despite patients' resistance, some regulation was ultimately required to provide security and safety. In 1918, two severely ill prisoners easily escaped from the Battambang hospital. Caught the next day, they were returned to the prison rather than the hospital. (The resident reasoned that their ability to cover more than sixty kilometers in less than thirty-six hours was indicative of reasonable health.) The chief of the Battambang hospital demanded that no more prisoners be admitted except in cases of surgery or moribund condition, due to the lax security in the hospital building.[112] Regulation was also required for the safety of other patients, who, depending on their condition, may have been vulnerable to violence. The Phnom Penh hospital sporadically witnessed cases of theft, attempted rape, knife attacks, and other dangerous and undesirable events.[113] To be effective, hospital administrators had to balance regulations with accommodation. As the following episode reveals, ineffective regulation of the hospital could create serious misunderstandings that highlighted the tensions in shifting definitions of life and death in a modern site of medical care.

## Laying Out the Dead

On September 27, 1934, at 4:45 PM, a French engineer in the colonial service named Charles Rigaud stormed into the local health office in Phnom Penh, demanding to see the health director. As was often the case, the health director, Henri-Charles Gérard, was out of his office, fulfilling one of the many bureaucratic obligations that were his responsibility. In the absence of Gérard, Rigaud expressed his displeasure to the unfortunate staff present. As one staff member stated in his deposition, "Mr. Rigaud, in his anger, uttered certain words that left me to understand that he was unhappy with the morgue." Upon leaving, the irate engineer left his calling card, on which he scribbled, "The medical chief is a bandit!"[114]

When he returned to his office, Gérard encountered his flustered staff with puzzlement, for he had never heard of Chief Engineer Rigaud, nor had he any idea what had caused his ire. But he was soon to learn a great deal more. Rigaud soon petitioned Résident Supérieur Achille Louis-Auguste Silvestre, a personal acquaintance, to open an investigation of the Mixed Hospital. He accused the hospital, Gérard, and the attending doctor of—among other things—malpractice, impropriety, and gross disrespect for the dead.

It seemed that Rigaud's sister-in-law, a Khmer woman named Oum Soy Bouk, had died in the hospital the previous day after emergency surgery to remove kidney stones. As was the Khmer custom, the family wanted the body cremated and the funeral to occur within the next day. When the family went to request the body at 4:30 that afternoon, some four hours after her death, they were informed that according to hospital regulations established to safeguard against epidemics, the body would be held for twenty-four hours unless the Phnom Penh mayor provided a special disposition for early release. Because the mayor's office was closed for the evening, the family decided to return as soon as possible after the twenty-four hour mark to collect the body. This was after the midday siesta the following day.[115] Rigaud and Oum's husband and brother came to the hospital at two p.m. the following day and found the morgue entrance locked with no attendants. Oum's husband had a cousin who worked in the hospital; his relatives soon persuaded him to open the morgue door. To their surprise and horror, the family found the woman's body heaped among several other corpses on a large stone table.

At their depositions, Oum's husband and cousin claimed no responsibility for Rigaud's actions, and indeed dared not impute any wrongdoing to the French establishment. Both employees of the French colonial government, they claimed to recognize that they had not complied with French rules and, though present and upset at the morgue that afternoon, were not participating in Rigaud's complaint.[116] The hospital's chief protested that regulations forbade the public from entering the public morgue for just such reasons. The hospital staff normally laid out the deceased in a private room adjacent to the common morgue for collection by the family after the twenty-four-hour holding period had passed. Rigaud, in response, called these medical regulations senseless. He fumed that even those with no medical training knew that "kidney stones are not an epidemic disease"; thus, why were they forced to wait at all to collect the body? After several weeks of investigations and depositions, RSC Silvestre, fighting chronic ill health, wrote wearily to Rigaud to give up his vendetta as a token of friendship and for the sake of French amity. It seems that the colonial administration had also taken steps to solve the problem, for Rigaud replied, "After two months have passed, I believe I understand that it's business as usual [*tout marche le mieux du monde*] at the Phnom Penh hospital, and that the natives closest to their unfortunate dead are satisfied with the honorable treatment reserved for [their loved ones'] mortal remains, left in heaps on the slabs of a sinister morgue, waiting to be buried. . . . Yet I am the one designated to go to Canossa! So, I will be going!"[117]

"For the sake of [their] friendship alone," he grudgingly withdrew the charge of "bandit" against Gérard.

Rigaud's reprimand by the colonial administration would indicate that the medical authorities and the colonial government again colluded against the patient.[118] However, unlike in the scandals involving vaccination, no person was actually physically harmed in this case.[119] Although the piling of corpses together shocked Khmer sensibilities and respect for the dead, it clearly also shocked the Frenchman's sensibilities. Regulations were in place to separate the medical view from the lay view of the body, both literally and figuratively. For the medical doctor, the lifeless body was a biological artifact to be stored until collected. For the lay person it was still a human being, unique and deserving of respect. The hospital held the body for twenty-four hours to decrease the risk of contagion; however, this regulation conflicted with the Khmer custom of rapid cremation and funeral rites for the deceased. The porousness of these regulations and the demands of Khmer custom ultimately resulted in the family stumbling upon the body in its "medical view."

Gérard was trapped between two opposing pressures: accommodation of indigenous customs and modernization of medical care. The perceived need for regulations, which were nonetheless often unenforceable, continued to create difficulties for the medical staff. During his stint as health director, Gérard had attempted to increase efficiency and improve regulations. These administrative attempts to rationalize the Phnom Penh hospital were met with a determined campaign by *La Presse Indochinoise* in the early 1930s to undo these new regulations and oust Gérard.[120] Indeed the campaign seems to have been in part successful, as he was replaced in late 1935 with a Dr. Louis Simon, whose arrival was heralded with some joy by the local newspapers.[121]

## Not Meeting in the Middle

Many were the barriers preventing the effective communication between Western doctor and Khmer patient. Attempts to overcome these obstacles were unorganized and sporadic. Baudoin's efforts to "localize" Western medicine by incorporating bits of Khmer traditional medicine were blocked by the medical administration. Gérard's efforts to move medicine in the other direction, by transforming the hospital into a rationalized, depersonalized site of care, were thwarted by the lay population. Unwilling to allow medicine to become more "native," the only way the French medical administration could move it in the same direction as it was moving in the metropole was to move the general population's attitudes. The AM had to transfer its focus from the one-to-one interaction between doctor and patient to the system-wide effort to influence the behavior and attitudes of populations. We will turn to this effort in the next chapter.

FOUR

# Social Medicine

We can trace three overlapping phases in the development of colonial medical services in Cambodia. The first and seemingly best option to improve native health was vaccination, a single shot that ideally prevented a variety of diseases in the long term. The vaccine simultaneously neutralized the threat of the disease to the European population and improved the health of the native population. The smallpox vaccination programs, which began seventeen years before the formation of the AM and continued through the colonial period, were the longest and most consistent effort undertaken by the Western medical services. Although smallpox vaccination was widely accepted in the countryside by the 1930s, smallpox epidemics were not eradicated during the colonial period. In part this was because those amenable to the procedure vaccinated too often, sometimes several times a year, whereas huge sectors of the population never attended a vaccination *séance*. In 1936 the Kompong Thom doctor would blame the incomplete success of vaccination on the indifference of the population, "save evidently during an epidemic, when on the contrary the vaccinator risks being trampled underfoot and suffocated from lack of breathable air, as they crowd in so numerous and dense . . . each attempting to be the first to arrive" (figures 4.1 and 4.2).[1] In 1937 the Kampot doctor estimated that at least 73 percent of the population was still susceptible to smallpox because most patients failed to understand the need for a booster shot every ten years, and others suffered under the misconception that smallpox was a child's disease, and thus vaccination was a child's treatment (figure 4.3).[2] For these and other reasons, more than half a century after the programs began, smallpox epidemics were still very much a thing of the present, much to the frustration of the colonial government.[3]

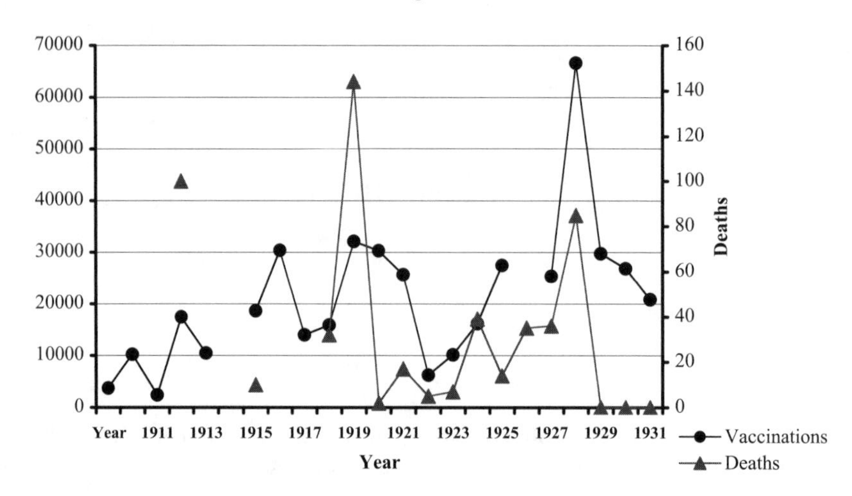


Figure 4.1. Annual AM smallpox vaccinations in relation to smallpox deaths registered in the province of Battambang.

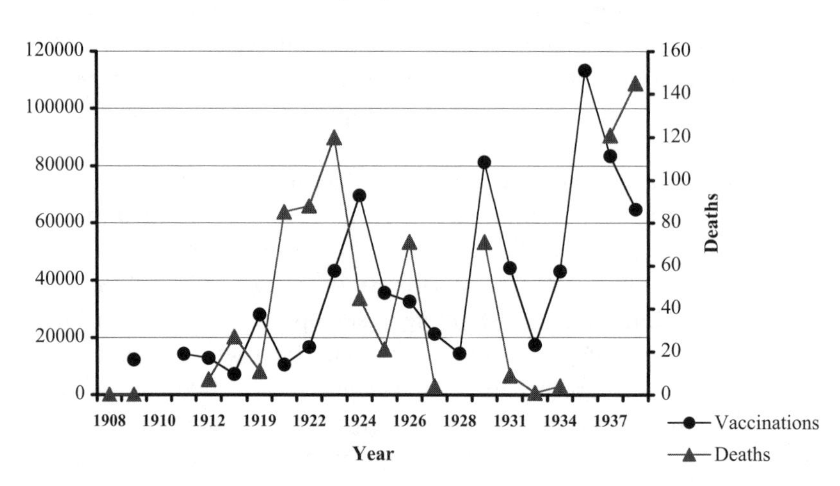


Figure 4.2. Annual AM smallpox vaccinations in relation to smallpox deaths registered in the province of Kompong Cham.

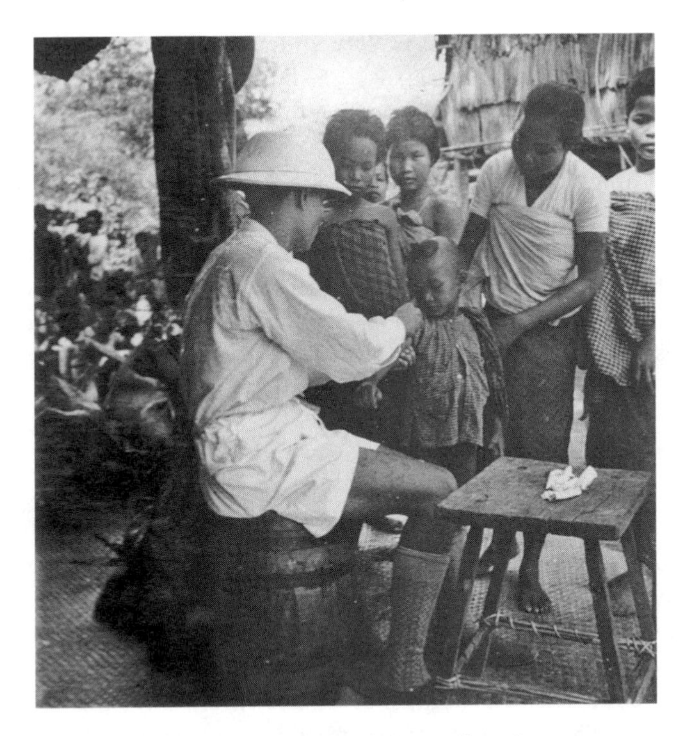

Figure 4.3. Routine vaccination in Khmer village, circa 1925.
File NAC RSC 30465. Courtesy of the National Archives of Cambodia.

The formation of the Assistance Médicale marks a second phase of the French medical program. At its creation, the AM's purpose was to expand the Western medical presence in the country, thereby bringing some tangible benefits of the French civilizing mission to conquered indigenous populations. Along with the ongoing vaccination programs, the medical service attempted to increase its presence in the indigenous setting with both practitioners and medical institutions. The Western doctor was an emissary of goodwill and humanitarian aid; his medical works were a key to converting those he treated into accepting the goodness of their "French protectors." However, the reach of the AM was constrained. Financially, it often ran on a deficit. The organization tightly budgeted medicines, supplies, buildings, and salaries. AM positions were in low demand: their prestige, job security, living conditions and posting stability were all unattractive. Further, a posting in Cambodia, which was seen as a frontier territory, was far less desirable than a posting in one of the three states of Vietnam. Ultimately, the sheer numerical weakness of Western-trained doctors and the budgetary tight-fistedness of colonial government made the vision of a blanket of French medicine over the entire population impossible.

In the 1920s and 1930s we can perceive another shift in the medical program away from this rather apostolic view of medical revelation imparted by the doctor to the patient, as the AM turned its focus away from the conversion of many by the medical actions of the doctor towards the attempts to change social health behaviors and environmental health problems through education, legal regulation, and public works. These efforts towards a population view of medical management generally fall under the rubric of public health, the main focus of this chapter.

## Public Health

Once medicine expands into the realm of public health or public hygiene, its implementation becomes a negotiation between collective good and individual freedom. It also involves conflicts in authority among the legislator, the hygienist, the laboratory scientist, and the practicing physician. This has been shown quite well in the context of nineteenth-century France, England, and America.[4] The negotiation between individual freedom and collective control has not been adequately studied in the colonial context. Too often it is simply assumed that the tyranny of public health and the tyranny of imperialism work neatly together. As one historian of medicine states, "Only in the colonies was the coordination between those who designed sewers, provided pure water, regulated frontiers and instigated quarantines, supervised and wiped out epidemics, and the public authorities, usually harmonious."[5] Yet, as we saw in chapter 2, this harmonious coordination was often illusory. Such a reading assumes that what is decreed is in fact carried out. However, as Penny Edwards observed in many of the political and educational reforms in colonial Cambodia, archival records of such decrees "express not events but plans and prescriptions of unfulfilled desires."[6] As late as 1936, the director of hygiene for Phnom Penh would wistfully note that private concerns had to be subordinated to public interests in order to have effective public hygiene—something that still had not occurred. He would observe of Indochina public health laws,

> [t]hese regulations, excellent in their conception, cannot have a genuinely useful effect, because under a possibly excessive worry about the protection of individual liberty, they prescribe a slow process that paralyses all immediate action. Because what in reality is Hygiene, if not the struggle for the col-

> lectivity! . . . To work for Hygiene is to encroach upon particular interests, to combat routine, prejudices, often ignorance, of folks who are at the base well meaning, but who have too much of a tendency to consider that "tout marche pour le mieux dans le meilleur des mondes"—and that Hygiene is only a pretext to perturb their tranquility and violate their liberty.[7]

Clearly, in the opinion of this French doctor and in the context of Cambodia, the laws passed by the colonial government were still shaped by pressures for individual liberty and against total social control in the colony. Like those of their metropolitan peers, the desires of colonial public health officials for such overarching controls were in large part unfulfilled, although for somewhat different reasons.

One could argue that the growth of public health movements in the colonies was *predominantly* about neutralizing the native population as a disease reservoir. However, without either the total annihilation of the indigenous population or the complete and perpetual quarantine of the European population, neutralizing the disease threat from the indigenous population still required a corresponding improvement in its health and strength, even if only for its more effective *mise en valeur*. The desire to strengthen the health of the lowest classes was, in both the metropolitan and colonial contexts, tied to the productivity of the nation. The great difference and ultimate hypocrisy of the colonial medical endeavor was that the economic harvest of this increased productivity was extracted from the colony to the metropole. Nonetheless, better health itself should not be extractable.

Historian of French medicine Erwin Ackerknecht argued that medical prophylaxis in the public health sphere was often an extension of politics, and that etiology, prophylactics, and politics were thus strongly correlated.[8] Historian Peter Baldwin argued against Ackerknecht, particularly in regard to national epidemic policies. Rather than evoking the neat triangle of disease, prevention, and politics, he contended that the main factor affecting where the pressure of prophylactic control was placed—on the individual body, on the group, or on the state—was the country's "geoepidemiological" location.[9] The colonial state would seem the ideal condition in which to examine whether or not epidemic control and state political traditions are in fact correlated, for only in the colony are the traditions of the controlling state as developed in one geoepidemiological location applied or not applied in another geoepidemiological location. As to the question of a geoepidemiological "learning curve," rather than serving as a laboratory for the metropole,[10] the colony benefited from the experience of Europe as its laboratory. The French experience with yellow fever and cholera epidemics of the eighteenth and nineteenth centuries in Europe[11] tempered French reactions to epidemic diseases in Southeast Asia. Initiatives taken by men on the ground were quickly brought into line with administrative and scientific understanding of proper epidemic control that often had been developed in Europe.

Consideration of the Ackerknecht-Baldwin debate is complicated in the colonial context by what historian J. S. Furnivall has termed the "abrogation of social will" by the subjugated population to the colonizing power.[12] This complication can be read in various ways, as public health involves an internalization of ideologies of epidemiological control. Some subaltern scholars claim that the colony is a unique situation in which "dominance is exercised without hegemony."[13] The ideologies of foreign rulers are often not adopted by the subjugated classes. If this formulation of subaltern scholarship includes matters of health and hygiene, then it would seem, regardless of "geoepidemiological location" or "politics, prophylactics, and etiology," that any colonial public health would be authoritarian and strongly repressive. Although colonial measures were repressive, particularly with epidemic control, how much *more* oppressive were they in the colonial context? Ultimately, an examination of the situation in Cambodia reveals the illusory nature of total control. Health Director Louis Simon observed in 1928 that the medical authorities did not have enough power to enforce their public health views. This lack of power allowed the "ignorant to infect the unsuspecting." He continued that due to this persistent ignorance, "a reason that I deplore . . . we are really insufficiently armed from the legal point of view. Because it must be said: if in Europe hygiene is a question of education and reason, in exotic countries it is a question of the police. It will be repressive or it will not be."[14] Simon's assessment was, however, incorrect. The ability of the colonial state to pass legislation may have been expansive, but ultimately that legislation was limited by indigenous cooperation. Surveillance and repression was not and could not be total.

On the other hand, studies have shown that in France itself, public health measures were often established well before the public accepted the reasoning behind them.[15] In many ways, these measures went hand in hand with elite ideology, and initially were enacted upon the middle and lower classes without their consent or their understanding of the justifications. Thus, the initial cooperation of the lower classes was minimal to nil in any context. "Domination without hegemony" is *not* unique to the colony. In light of this, the question of "prophylactic ideology" becomes an exercise in elite rationale rather than any meaningful analysis of events or policies as they were enacted. Legislators often enacted policies without the consent or understanding of the affected populations in both the colony and the metropole. In both contexts, the policies or the practices they engendered were then revised in negotiation or compromise with the affected populations. These revisions were, of course, dependent on context. Further, many of the regulations that seemed particularly harsh on paper were unpracticed and indeed impracticable in reality. The rhetoric behind the laws gives us little understanding of their actual effect. A decree did not mean control or even modification of the existing situation. Thus, only in examining the actual effect of policy—not simply the rhetoric behind it—do we begin to understand the difference between these processes in the colony and in France, and to understand what is truly colonial about "colonial medicine."

If initially enacted from the top down, public health eventually had to involve some public cooperation. This cooperation was dependent on some general understanding consistent with the measures the French implemented. The remainder of this chapter will look at French epidemic control measures and systematic attempts to bring indigenous peoples in line with these measures, including efforts to create this elusive public cooperation, and thus public understanding. We will briefly review public health policy in urban centers, particularly as it dealt with epidemic control, before moving to what could more accurately be termed public health outreach, or attempts at populist appeal by the French in the Cambodian countryside. We will examine how epidemic control measures gradually expanded into more general public health regulations, and how this expansion was driven not by circumstances in Cambodia, but by the social and political climate in neighboring Vietnam and distant Europe.

## Managing Epidemics

Southeast Asia has been identified as a crossroads of commerce and cultures. This openness, ironically, has in fact been labeled as one of the characteristics that binds it as a coherent region.[16] Already known as a nexus of trade, it saw an increase in international traffic and commerce during the nineteenth century with British, Dutch, and French colonial efforts in Southeast Asia.[17] This growth in trade led to increased flow of diseases across borders—what one anthropologist has called the "microbial unification" of the globe.[18] Thus the medical administration in Indochina had an early interest in epidemic diseases. Epidemics disrupted the commercial activity of the colony, and commercial activity was of prime importance to each colonial government, including Indochina. Early public health efforts in Cambodia, and indeed all of Southeast Asia, were tied to efforts to contain these economically disruptive diseases. This is not unique to the colonial context. Public health as a field developed in intimate relation with economic and political concerns with epidemic control. As in Europe, epidemic control—which would become inextricably linked to public health—predated organized efforts at public hygiene.

In 1902 the minister of colonies decreed a list of epidemic maladies requiring open declaration (*dèpistage*) to authorities.[19] In 1911 this decree was expanded to state explicitly that doctors could not invoke professional medical secrecy to protect patients. The diseases, listed numerically, were sometimes referred to in shorthand. For example, cholera was often referred to in telegrams as disease seven, and leprosy as disease fifteen.[20] The measures against epidemic diseases, many worked out somewhat haphazardly by doctors or administrators during epidemics, had become standardized by the 1910s. For smallpox epidemics, the 1913 decree of GGI Albert Sarraut emphasized continued vaccination runs. For other major epidemic diseases, particularly plague, cholera, and recurrent fever (*fièvre recurrente*), declaration, containment, and disinfection were the main prescriptions.[21] For the most part these epidemic measures were city-focused; Phnom Penh and Kampot were major centers for epidemic control. Epidemic measures also varied by disease. The plague and cholera both called for isolation of the sick, burial under lime or quick incineration of the dead, and disinfection of homes and material goods. Preventative inoculations, originally mandatory for the plague in 1907 and for cholera immediately after World War I, quickly became facultative for both diseases due to public resistance.[22] In the provinces, few of these epidemic measures were enforced. For other epidemic and pseudo-epidemic diseases on the list, the health services undertook other measures.

Malaria, one of the most debilitating afflictions in Cambodia, was one such disease. Although more accurately endemic than epidemic, the colonial government classified it as an epidemic disease. Throughout the colonial period it decimated the population of the Cambodian countryside and represented a serious obstacle to French efforts at economic development. The administration expended considerable resources researching means of combating it.[23] At the end of the nineteenth century, the only effective prophylactic against malaria was quinine.[24] From its inception, the AM freely distributed quinine to regional governors; initially, regional heads often failed to redistribute it to the local population, seeing no purpose in it.[25] On

December 4, 1909, the GGI created the Quinine d'Etat Service, establishing quinine depots for the general population in the five states of Indochina.[26] The AM freely distributed quinine to these depots, which were strategically placed in malarial regions for public access. The *mesroks*,[27] the population at large, and even secondary distributors were encouraged to obtain the medication and redistribute it.[28] After the international scientific community accepted the role of the anopheles mosquito as the vector of the malaria plasmodium,[29] the medical services also advised the indigenous population to use mosquito nets—advice that went largely unheeded. Although some AM doctors attributed the lack of mosquito net use to local recalcitrance, Dr. Bernard Menaut would observe in 1923 that poverty prevented its purchase and use: "Any time that a Cambodian is able to buy a mosquito net, he will, without any consideration of prophylaxis pushing him to this purchase; all creatures seek tranquility and the Cambodian strongly appreciates the tranquility of his nights assured by a mosquito net. . . The usage of the mosquito net will progress in parallel to the prosperity of the country."[30]

The colonial administration also discussed several public works projects to sanitize unhealthy terrains via water drainage and brush clearance, particularly in parts of Vietnam.[31] In Cambodia, due to the extensive periodic floods and relatively late colonial interest in agricultural development of malarial regions, few public works programs were planned. By 1938 the French military began employing the new synthetic drugs quinacrine and premaline against malaria in northeastern Cambodia.[32] In the opinion of the Haut Chhlong doctor, quinine had almost no positive effect against the disease, while this new "chemotherapy" was a dramatic improvement.[33]

## European Decline in the Khmer Brush

Malaria, as discussed above, and leprosy (which was approached in an even more puzzling way by the administration; I discuss that disease in its own chapter) were endemic diseases that received exceptional early attention by Western medicine. The programs to control them represent some of the first organized efforts at public health outreach. The following section focuses on isolated and less organized efforts and their gradual expansion to reflect an official administrative stance towards "social medicine." To understand these policy changes, we will trace the underlying social currents that impelled their development. These changes are linked not only to the imperial concept of *mise en valeur*, but also to greater societal anxieties, misplaced onto Khmer society—a phantasmic fear of French decline lurking in the

Khmer brush. Further, this fear resonated in the context of contemporary colonial scholarship with existing tropes of Khmer degeneration from the Angkorean period, and French efforts to revive a failing society.

National decline had been an obsession in France since that country's resounding defeat in the Franco-Prussian war. The concept of what historian Robert Nye has termed "the medical model of cultural crisis" was developed during the belle époque to explain perceived French national degeneration. Degeneration explained the decrease in population and political/military strength relative to other European nations, as well as the increased alcoholism, suicide, venereal disease, and pornography in this period.[34] Nye locates this angst in the "altered politics of a defeated and divided nation and the bleak ruminations that followed."[35] While originating in the losses of the Franco-Prussian War and the bloody Communard period that followed, this fear of national decline intensified after France's costly World War I victory.

In many Western nations, the interwar period marked the height of the infusion of the biomedical perspective into social management, especially as represented by the eugenics movement—the effort to improve society through selective breeding of human beings—which reached its apotheosis in the horrors of Nazi Germany. It is difficult to untangle the knot of eugenic ideas that became incorporated into the increasingly powerful hygienic movement. Historians of science tie Progressive-era sociocultural shifts in the U.S. context to the growth of sanitary sciences; in this view, the growing belief that science should become involved in the management of all aspects of society facilitated public health as a field. Similar trends can be observed in all of the major countries of the West. The unique slant of eugenics in France originated from the country's perceived crisis of national depopulation. In a country where more than 10 percent of the population had been killed in World War I and another one million were permanently maimed,[36] negative eugenics—prevention of "undesirable births" through sterilization, abortion, and contraceptives—was almost wholly blocked.[37] For similar reasons, eugenics in France was unlike other eugenics movements in being strongly tied to a national natalist movement.[38] However, there was at least some indication of a racialist component in the French eugenic movement in Cambodia. For instance, on April 6, 1938, the secretary general of the Buddhist Institute in Phnom Penh forwarded to the GGI a detailed transcript of an interview conducted between journalists and the French minister of public health on the as-yet-unrealized "health card" to protect the French race. In a hopeful note accompanying this transcript, the secretary general observed, "In Indochina, where there are not the same prejudices to conquer as in France, the creation of these cards would admirably complete the grand oeuvre of the preservation of the races undertaken by the French administration in this Colony, with all of its hospitals, dispensaries, and maternities."[39] An administrator in Saigon sent back a polite memo that the issue would be studied.[40] However, the GGI seemed to have little serious interest in the issue of a race or health card, although we can identify concern with race in the changes that occurred in naturalization laws during the 1920s and 1930s.[41]

The growth of sanitary sciences meant a switch in the AM's focus from acute medical intervention to efforts at changed social behaviors—to prevention over cure. These changes mirrored the changing political alignments in France. The increased focus on public hygiene and public education in Cambodia began after World War I and reached its peak in the mid-1930s concomitant with the ascension of the Popular Front in metropolitan France.[42] Thus various colonial decrees proliferated, expanding public hygiene and social medicine sectors. In April 1918, the GGI split the health services in Cambodia into a medical service and a hygiene service. The hygiene service was intended to be mobile in the countryside, but its functions quickly became absorbed by the needs of the Phnom Penh municipality.[43] In 1922 a GGI directive further created an experimental hygiene sector in each of the states of Indochina, modeled on the previously attempted hygiene sector in Thanh-Hoa (Annam, Vietnam). These hygiene sectors were organized under the explicit goal of "hygienic education of the natives" at the village level. Their plan of action included creating accurate malarial indexes for each region,[44] educating village notables on the means of ameliorating diseases in each village, and extending smallpox vaccination and free quinine distributions. The GGI stated with some confidence that after this fundamental groundwork had been laid, "it will be easy to bring the natives to understand the necessity of the work to be done, [work] which can only be done with a shared mutual understanding and without any coercion."[45] This "work to be done" was both public works (building wells, draining swamps, and so forth) and peer education, changed "unsanitary habits," and other "improvements" to the hygienic setting of the indigenous population.

In Cambodia, the health director created this experimental hygiene sector in Kompong Cham, tapping Dr. Menaut as its head.[46] Under his energetic direction, the hygiene sector performed impressively in its first year of operation.[47] However, its work soon tapered off. By the time Menaut was promoted to health director in 1927, it had become only a ghost in the administrative record, with no effective programmatic functioning.[48] Along with the creation of this sector, the earlier health director had made other changes in 1922, most notably adding a section on the "mobile hygiene and prophylaxis" service in the AM's annual medical report for Cambodia.[49] In reality, the "mobile" part of the service was often neglected in favor of the demands of Phnom Penh city.[50] By the following year, the Phnom Penh municipal doctor was also named the head of this mobile hygiene service. The added title did not mean additional duties; by this point the service had contracted essentially to urban epidemiological monitoring, a task that the municipal doctor already performed.[51] In 1923, the Ministry of Colonies requested information on all social assistance and social hygiene programs in Cambodia. Menaut, the report's author, observed the lack of most types of social medicine in the country, particularly in comparison to Vietnam. This, he believed, was due to the country's underdeveloped economic state, as "the provinces of Cambodia do not have the wealth of the provinces of Cochinchina."[52]

## Healthy, Happy, and Politically Content

A noticeable spike in administrative decrees on the sanitary improvement of the indigenous population occured in the beginning of the 1930s. This flurry of legislation could be attributed to an anxiety over the recent unrest in Vietnam, most notably the formation of the Nghe-Tinh Soviets, a rebellion that unseated large swaths of the French colonial government in central Vietnam for more than eight months in 1930 and –1931.[53] Modern scholars point to the role of the Great Depression and the serious famine in parts of Vietnam in 1930 as motivating factors for the rebellion.[54] The link between colonial fear of unrest and interest in native health seems clear in a two-page questionnaire sent out by the inspector general to all medical directors in March 1930. The questionnaire posed questions such as, "For the native of the least fortunate classes, is he undernourished? If he is undernourished, is this constant, seasonal, or accidental?" and "Is the native sufficiently clothed to resist intemperate weather, variations in temperature?" The final and most important question was, "Have you seen any tendency toward amelioration of these conditions?"[55] Again, the seemingly ubiquitous Menaut answered the questionnaire for Cambodia with a nineteen-page report on food, clothing, and lodging for the "natives of the poor classes." Interestingly, he divided the populations into three ecological lifestyle categories (fields, riverbanks, and forests) and separately described the populations of each. He concluded that for the most part, the constant warm temperatures in Cambodia made the flimsy homes and scant clothing sufficient protection for their owners. However, most of the population was generally undernourished even in plentiful periods, leaving them easily susceptible to disease and starvation in periods of want. As to improvement, Menaut would note that in his twenty-four years of colonial service in the country, great progress in material well-being had been made amongst the upper class, little in the middle class, and absolutely none in the poor classes. He would conclude: "If it is absolutely true that one needs to eat to live, it is no less true that one needs to have a certain alimentary hygiene to live in good health. This alimentary hygiene is nonexistent in the poorer classes [and] it has more influence on the mortality that one observes than undernourishment or periods of food shortage."[56] In other words, general inequality *and* health inequality had increased under French colonialism, with most of the population still receiving no benefit.

Ultimately, the link between actual unrest in Vietnam, potential unrest in Cambodia, and French administrative concern for material well-being of the indigenous population was tenuous.[57] The questionnaire was disseminated in March 1930, two months *before* the Nghe-Tinh movement erupted.[58] Sanitary measures were not undertaken merely to forestall social unrest—or if they were, they displayed rare foresight rather than learned hindsight. It is more likely that the many social, political, and economic anxieties of metropolitan medicine had crept into the consciousness of the colonial medical administration with the encroaching global depression and social unrest in other parts of the world. More plausibly linked to the idea of "learned hindsight" are the series of hygienic reforms the colonial government undertook in the year following the Nghe-Tinh Soviets. In 1931 the health services attempted again to expand the hygiene sector, ordering the creation of a distinct service of hygiene and preventative medicine in each state.[59] Both services—medical and hygiene—still operated under the control of the Cambodian Health Service, but each would have its own budget and personnel. The underlying purpose of this split was to assign devoted and separate personnel to each service; doctors were thus far too occupied with their hospital functions to worry about preventative medicine, despite previous calls to expand social hygiene. However, the inspector general's directive was already old hat in Cambodia. As we will recall—and as the governor general himself did not—the GGI had already decreed a split in the Cambodian medical and hygienic services in 1918.[60] In 1931, RSC Lavit informed the governor general that the previously decreed service was already functioning in Cambodia. He would add, somewhat nonsensically, that *aside from the continuing vacancies in the service due to the lack of personnel and budget*, this hygiene service "always provided excellent results."[61] In other words, excepting the inability of the service to actually provide health care, it worked well. In the same year, the health director attempted to expand the hygiene sector as implemented in Kompong Cham in 1922 to the provinces of two other provinces. However, this plan came to a halt when the AM could not provide more personnel for the task. The inspector general promised a new military graduate for the service, but none was forthcoming.[62] On July 13, 1932, a GGI decree explicitly split the civil and military sanitary services. Once again this decree emphasized outreach: the doctor was to leave his medical center and go to the villages, medicine was to be distributed locally, and natives were to be encouraged to start their own social assistance programs and to "care for each other." The decree also stressed the need for more demographic and medical information, in order to have a real assessment of disease levels in the countryside.[63] Three years later, in 1935, Louis Rollin distributed another ministerial circular echoing this official stance.[64] In the middle of this decade, infant mortality would become one of the key markers of health status for nations around the world. Thus, Rollin's circular emphasized that doctors should focus on decreasing infant mortality. Not coincidentally, the following year, the annual medical reports began to include tables of comparative infant mortality by nation.[65] The status of a population's infants would become the crux of population status, the representative metric for a society's health. It remains a key metric of population health today.

In one of the last medical efforts before the Vichy period, the health administration created a new mobile medical service in 1938. The automobile marked the novelty of this service as opposed to the various mobile services that had come before. This service's automobile had an attached trailer, allowing patients to be treated at the vehicle itself. It was not quite a modern ambulance, but it was a considerable improvement. This service was also charged with demographic and epidemiological tasks such as mapping village locations, population counts, and disease tallies by region.[66] With the outbreak of World War II, however, this service was discontinued; it was ultimately too transient to effect real change.

While these decrees reveal how the colonial state was discursively reconfiguring itself and its subject populations through the language of medical intervention, the impact of these decrees on the local populations was likely insignificant. If we take the amount of discussion and legislation on social hygiene as an indication, the 1920s and 1930s were decades of tremendous change in the French health services in Cambodia. In reality, the changes were less significant. Hygiene sectors quickly disappeared, mobile services became absorbed into preexisting medical services, and administrative restructurings were for the most part transient. The next sections examine how these calls for social medicine were specifically enacted and interpreted in Cambodia. We first detail a few propaganda pieces circulated by the administration, before turning to educational efforts. Because the educational structure in Cambodia was so rudimentary, the discussion on hygiene education in schools is brief, with more attention given to the education of entire villages in informational hygiene meetings. A careful analysis of the perspectives of these local actors (doctors, patients, and bureaucrats) reveals the subtle evolution of medical significations and practices.

## Plural Etiologies

Key to Western ideas of public health is the term *contagion*. The concept of passing a disease from one person to another was not unknown in the context of Khmer traditional beliefs, but it was very specific to certain diseases: smallpox and some mental illnesses, for example. It is instructive to compare this to the stance of the metropolitan French medical establishment half a century before the AM's formation. In 1846 the majority of the French Academy of Medicine voted with the anticontagionist Antoine Clot-Bey on the question of plague and quarantines, advising the French Ministry of Agriculture and Commerce that the plague was not contagious but rather was spontaneously generated under favorable conditions, with miasmas playing a large part in its production.[67] Contagion was not rejected because it was new, since its roots traced back to the Hellenic period. On the contrary, by the time of this 1846 vote the scientific community largely viewed the theory of contagion as old-fashioned and outmoded in comparison to the environmentalist stance developed at the beginning of the century. By 1900, the wheel would turn full circle.

Contagion had undergone a radical transformation in the West in the second half of the nineteenth century. With the work of Pasteur and other microbiologists, the "transferability" of diseases had moved from a focus on the soil (the patient) to the seed (the germ). Disease etiology was transformed from an inchoate force caused by factors ranging from moral failing and environmental miasma to a discrete physical entity, as the increasing incontrovertible proof of the microbe made a "*guerre au bacille*" possible.[68] This transformation was still in process among the doctors and researchers in France as well as its colonies during the time of our study. Further, among the European lay population, there was also a lag in the transformation of the conceptual understanding of germ theory and contagion. Often the public's acquiescence to new methods of public health predated their understanding of the reasoning behind these methods.[69] One could argue even today that the gap between the scientific and lay understanding of germ theory is substantial.

The term *chhlong* is perhaps the closest Khmer translation of "contagion." It literally means "to cross," but it can also refer to the transfer of a sickness from one person to the next.[70] This definition could be seen as close to the pre-germ theory definition of contagion in French or English. In the era of germ theory, the key difference in the two meanings of contagion involves the concept of "germ." The germ (in its later guise, "microbe"), an entity invisible to the naked eye, allowed a one-to-one correspondence between a disease and its cause. This concreteness minimized disease as a moral failing of individuals and recreated it as a process of physical contact between people. Certainly, Khmers recognized in some instances that physical contact with a sick person caused some illnesses, but only very specific illnesses. Again it is instructive to compare this with the famous 1846 vote of the French Academy of Medicine, when even the most fervent anticontagionists admitted that specific diseases (syphilis, smallpox, gonorrhea) were contagious. They refused to apply this theory, however, to the wider gamut of diseases such as plague, yellow fever, and cholera. For such diseases, telluric etiologies and hereditary essentialism were thought to play a key role.[71] In the Cambodian context, the gap between the native and Western notions of contagion was compounded with the newness of a disease such as the plague, as well as with the new severity of other highly contagious diseases such as cholera and smallpox.[72] One of the key points of confusion in giving physicality to the disease cause was the notion of carrier. To insist that an apparently healthy person (healthy carrier or transition carrier) was a harbinger of disease, or that an animal (intermediate carrier, or vector) had spread the disease from another species to a human being—these concepts could appear more mystical than existing indigenous beliefs. Further, to argue that the physical emissions of a human being were somehow imbued with disease-causing properties was a conceptual leap.[73] The relative newness of bacteriological theory in medical practice during the colonial era is not obvious when one reads the barrage of administrative comments imbued with scientifically-informed scorn over the "backwardness" of the colonial subject. In a typical instance, one deeply unhappy doctor in Takeo would identify native "lack of concern" with contagion as an indigenous character flaw preventing patients from enjoying the benefits of modern medicine.

> [N]othing is of less concern to local customs than the worry of contagion, respect for isolation and instructions for their protection, no one is less apt than our infirmiers, not only in encouraging sanitary defense measures, but also in maintaining these measures with a relatively simple rigor, no one is less apt than the [village] chiefs to give these measures their support and the consecration of authority. . . . [Where] there is a smallpox or cholera patient in a village; everywhere that I have made a house call, they sit on the patient's bed, they feel him, touch him, wipe him, etc. etc. without regard for his sores or pustules, without the least worry about all the suppurations, emissions, secretions that he is producing.[74]

This scientific scorn was not only an instance of authoritarian disgust over the barbarism of its authorized subject, but also representative of the then-recent conversion of the medical profession to germ theory, and its attempts to distance itself from its own "less" enlightened past. This tension is apparent when we review attempts by specialists to educate their subjects, "barbaric" or otherwise, on the rational definition of disease.

## Advertisement

Early French efforts at what could be called population education or propaganda began with epidemics, but eventually expanded to other aspects of indigenous health. These endeavors had to contend not only with the language barriers of a multilingual colony (Khmer, Vietnamese, Chinese, and Cham were the major languages) and the attendant problems of mutual incomprehension, but more seriously with the differing definitions surrounding disease, sickness, and health that could not be bridged by simple translation. These definitions were further confounded by the growing gap in French between the specialized language of scientific medicine and the quotidian language for health and illness. Some early efforts at publicity for epidemic control illustrate this difficulty.

Cambodia's first plague epidemic occurred in 1907. The RSC requested that the Khmer Council of Ministers produce an announcement educating the public about this new disease. The original draft of the poster that the council submitted for French approval referred to the disease as *rook puh* and emphasized the need for doctor's visits and disclosure to public health officials.[75] The word *rook* was intended as a translation of "microbe," but a more accurate translation would have been "disease" or "illness." A Khmer word did not yet exist for a biological entity invisible to the human eye, a concept that would indeed have been alien to indigenous conceptions of illness at the time.[76] Further, the term *puh* meant "poison"—or, in the context of injury, "swelling." Thus, the original announcement presented the plague as the "swelling disease." It was an odd linguistic coincidence that the word *puh* not only sounded similar to the Khmer pronunciation (la peh) of the French word for the plague (*la peste*), but also signified some of the disease symptoms. The medical services feared that referring to the disease as *rook puh* would suggest that it was a variation of an existing illness, thus leading to therapeutic confusion. The bubonic plague, which was at this historical moment being transformed from a clinical symptomology to a laboratory-confirmed microbe under its French name, could not be semiotically linked to other diseases when given a Khmer name. The final version of the poster dropped the word *rook* and simply referenced the disease in French transliteration as *la peh*.[77] While translation is usually a search for equivalencies, in this case it was also a search for difference. That is, the French medical services had to carefully distance existing linguistic associations with the disease. The plague had to be named in such a way as to separate it from any preexisting indigenous connotations, even if those connotations were appropriate to its clinical symptoms. The French colonial government, both source of and solution to the plague (and neither one quite completely), would also control its naming.[78] However, this control would also be incomplete. A decade later, in 1919, a French provincial administrator distributed a poster in Khmer on the destruction of rats in order to fight the ongoing plague epidemic. The poster referred to the plague as *chomgnu puh*, or the "swelling sickness," rather than *la peh*.[79] The Khmer term *puh* continued to survive alongside the French term *peh*, even among the French administration.

The idea of the "microbe" was broached again by the AM in the late 1920s, but this time the administration made some effort to explain the concept in "local" terms. As one French doctor explained, "the head of the rice farmer is not made for scientific exposés, yet one can explain important practical notions by drawing from his observations of daily life and reasoning with him by analogy and image."[80] In an attempt to fight the rampant levels of conjunctivitis in the countryside, the health services distributed an informational poster in Khmer and quoc-ngu (the romanized Vietnamese alphabet) to roughly 1300 *khums* (clusters of hamlets or *phums*) in the late 1920s.[81] The poster's language was indeed basic; it explained bacterial infection as a sort of parasitic infection writ small. Written in plain Khmer, it decried the traditional "ugly habit" of grinding up dried frogs and using them as a paste on injured eyes. The health services administration demonstrated relative creativity in its translation, replacing the concept of bacteria with

Figure 4.4. Poster advertising quinine usage. Courtesy of the National Archives of Cambodia.

that of a "very small worm." As the poster explained, this "ugly habit" was very dangerous because in the threads of the frog's intestines lived "a very small worm" that climbed up under the eyelid and made the sickness worse, possibly leading to blindness. In rather strong language, the poster urged natives to abolish this "disgusting" habit and show a little more discernment over such practices.[82] The poster actually made no mention of available French medicines or visits to the French doctor; its sole purpose was to discourage a traditional practice in Khmer society.[83] It thus represented a colonial attempt to educate the Khmer populace outside the realm of direct doctor-patient interaction.

In a society that was largely illiterate, the value of these posters in the countryside was likely limited by the willingness of the *mekhum*[84] to translate their message to villagers. The use of visual images was surprisingly rare. In one of the few exceptions, the health services distributed a poster on malaria and quinine. This poster not only is unusual in incorporating a visual image, but is one of the few pieces of extant propaganda that relies *predominantly* on the visual image rather than the written word. It proclaims that "kee-neeng" (quinine transliterated) is a very powerful medicine. Above an image of two sick Khmer men shivering in blankets, the caption reads, "A mosquito bit him and made him feverish." Over the image of a dancing man holding a huge quinine pill, the caption reads, "He swallowed the quinine and was healed of fever." Under the center caption urging sick individuals to request quinine, the "strongest medicine of all," from authorities, sits a huge mosquito (figure 4.4).[85] This poster seems to have been reproduced several times by hand, but never mechanically reproduced. Thus it is doubtful that it had a wide distribution. Even today, many country dwellers in Cambodia still do not link malaria to mosquitoes.[86]

## A Failure to Communicate

Beyond fliers and posters, the administration also made more general efforts to educate the public on health. Besides the Baudoin brochure (previous chapter), Dr. Adrien Pannetier in 1906 wrote a small health manual that the government redistributed in 1913 to a limited number of schools.[87] This was an exceptional and isolated effort, despite the GGI's 1907 directive which urged health education in schools, general outreach, and all branches of medicine to work together to enable "colonists to flourish on less deadly grounds and to provide to the natives a guarantee of well-being and health that they have not yet encountered."[88] Due to the rudimentary nature of education in Cambodia, teaching public hygiene in schools was of limited effectiveness. Furthermore, the disproportionately low school attendance of females, who were often the initial caregivers to the sick, also served to limit the reach of health education. For example, a circular in April 1925 decreed the establishment of a physiological file for students of all public *écoles*. The district doctors were responsible for maintaining these files by periodically visiting the *écoles publiques* and performing physical examinations on the students.[89] No records exist that such files were actually kept, although a note in the 1934 annual report claimed that four schools were regularly "surveyed."[90] As for other aspects of the 1907 directive, including the call for all aspects of medicine to work together for the greater good, they seem to have been largely disregarded. In 1910 one of the provincial doctors would write in disgust of these instructions,

> The governor general, who sees things from on high, wants constant collaboration between the two elements [of the health services], technical and administrative "Ce sont ses propres paroles". . . . He released a circular that

> dates back to 1907, and which we have before our eyes, soliciting the formation of provincial hygiene commissions[,] but oh tempora mores! Where are the idyllic virtues of the first ages of Jean-Jacques! This circular is taken note of, it is surely registered, without forgetting either periods or commas. But it is hastily ignored. There is no need to be a great scholar to know why. . . . To apply an ideal system of sanitary defense is impossible in Europe: for even greater reason here where one must contend against the elements, periodic inundations which flood 3/4 of Cambodia for five to six months of the year, a lax social framework filled with holes and insufficient education, a people unaccustomed to discipline, too scattered, too distant from each other, etc.[91]

Indeed, although the GGI may have recognized somewhat early on that the mere presence of the French doctor was not going to move the indigenous population towards Western medicine, few organizational efforts were made by the health services in other regards. Not until World War I and later did the health director begin to encourage doctors to educate their patients whenever they were given the opportunity. In Prey Veng in 1921, the *infirmiers* were told to encourage inhabitants to "observe elementary rules of hygiene" when they procured government-distributed medicines.[92] Further, the Prey Veng *résident* also informed the village and regional chiefs that the "best" of those who observed and encouraged such behavior would be recommended for an honorific title.[93] In Cambodia more widely, doctors began providing educational sessions around 1919 while on tour in the countryside. These meetings often were unproductive, however, as the sometimes off-putting nature of the interaction between the Khmers and French as well as the perceived disrespect for preexisting customs and habits created further misunderstandings. Further, not only "words" but ideas and the way they were comprehended were not effectively conveyed, in either direction.

Ernest Gellner, in his classic work on nations and nationalism, argued that one of the most striking intellectual differences between industrial society and the premodern (or "agro-literate" society) was the coexistence within the premodern world of "multiple, not properly united, but hierarchically related sub-worlds, and the existence of special privileged facts, sacralized and exempt from ordinary treatment." In contrast, we moderns locate "all facts . . . within a single continuous logical space"; to do so is part of what we conceive as rational. However,

> in a traditional social order, the languages of the hunt, of harvesting, of various rituals, of the council room, of the kitchen or harem, all form autonomous

> systems: to conjoin statements drawn from these various disparate fields, to prove for inconsistencies between them, to try to unify them all, this would be a social solecism or worse, probably blasphemy or impiety, and the very endeavor would be unintelligible.[94]

This framework is certainly useful for understanding some of the language problems between the French doctor and the indigenous villager. The tension between French and other, scientific and superstitious, mundane and mystical existed much more strongly for the colonizers than for the colonized. French administrators viewed the absence of this tension in Khmer reactions as indifference or superstitious irrationality. At times it seemed to certain Frenchmen that Khmer incomprehension was entirely willful. The seemingly elusive "logical consistency" that the French sought in Khmer reactions often discouraged concerted French attempts to propagandize their medicine. Khmers, in contrast, must have found the inflexible logic of the Frenchmen odd and unnecessary.

One detailed example exists of a 1927 village health education meeting performed by a Khmer *médecin*, Khuon Kimsann. The report's language is fascinating both for its description of the doctor's role in the village and for the tension that this *médecin* clearly faced in aligning his identity more closely with the French than the native villagers. Written to his French superiors on his work in educating natives on the true causes of malaria, the report consists of several pages of first-person description of his talk—a great deal of which covers his own imperviousness to malaria due to his greater knowledge. After describing to the villagers the life cycle of the malaria plasmodium (the Khmer word he used is unclear, as the report is written in French) and the need for everyone to use a mosquito net, he concluded the talk thus:

> If you rigorously observe these prophylactic rules, you will no longer have anything to fear, you can break from all ancestral recommendations on this subject, and also pass over the head of the "nakta" whom you fear the most. . . . Like you, I am Cambodian. Thus, I have no interest in leading you down the wrong path. All that I have just told you is nothing but the entire truth, based on in-depth studies and observations that I myself have made during the course of my career. Thus, you can believe it. . . . Our Medicine, if it is Medicine, is still very rudimentary. One only needs to compare it to European Medicine, whose progress daily moves forward with giant strides. Everything that is a mystery for you is well-known in Europe. You are still very backwards in this regard, and it is about time that you relieve yourself of all

> these superstitious beliefs that only create prejudices, therein preventing you from profiting from the progress of science.[95]

The *neak ta* ("nakta"), the ancestor spirit of each village, was seen as a powerful protector or destructor genie, depending on his mood. In Khmer culture, the head is sacred. To "pass over the head" of anyone is one of the highest forms of insult, and to suggest that the Khmer "pass over the head" of their *neak ta* would indeed be sacrilegious. Whether this speech was a close translation of the Khmer speech or revised for the eyes of his French superiors, the writer was clearly attempting to separate himself from the backwardness of his patients, and to do so in much stronger language than the French practitioner had used.

While these men were meant to bridge the gap between cultures, bringing together cultural identities was not such a simple task. Cambodian scholar Penny Edwards observed that the Khmer civil servants of the French colonial administration were dynamic agents of change, but would negotiate these changes on their own terms. She also suggested that some Khmer leaders became "more Khmer" in the face of French efforts to reform Cambodian culture.[96] In a different vein, an ethnologist researching modern public health programs observed that traditional medical practitioners working for development agencies today often exist in a framework in which the attempts to translate the "universalism" of development discourse into regional areas of action actually decontextualize the local context and make it irrelevant. In the process of facilitating program goals these workers, rather than serving their intended role as interlocutors, would "side-switch" and reject their original culture in their work.[97] It certainly seemed that several indigenous *médecins* in Cambodia chose to "side-switch" rather than retrench within a Khmer identity. In the 1920s, Pannetier noted another (unnamed) Indochinese *médecin* whom Khmers avoided due to his refusal to use his native Khmer language, speaking to patients only through an interpreter and in what he perceived as the correct French manner: the distant formal form of address.[98] Kimsann, who entered the service after Pannetier's writing, would likely have had the same effect on the indigenous population, despite his self-congratulatory note in his 1927 medical report.[99]

In 1919, Kompong Cham medical chief Honoré-Matharin le Nestour made a trip to a village suffering from a cholera outbreak. While there, he not only gave consultations and performed several vaccinations but also called the villagers together to speak to them about public hygiene and the measures to take against cholera. During the talk he was much encouraged by the villagers' attentiveness. They repeatedly assured him of their comprehension and total agreement with his points, such as the necessity of boiling water before drinking it, confining the contagious, and disinfecting contaminated property. Satisfied that he had achieved his end, Nestour was approached immediately after his talk by one of the villagers. The attendee, very respectful of the doctor's speech, asked Nestour's permission to organize a ceremony to float a miniature pagoda with an offering of pork onto the river to appease the divinities causing the cholera epidemic.[100] Nestour's sense of accomplishment was quickly deflated. The attendee's pride in his collaboration was doubtless diminished by the doctor's reaction. The problem was not simply one of language. The doctor most likely had an interpreter with him; most French doctors were equipped with an interpreter or aide when traveling in the bush. It seems that the villagers essentially saw no inherent contradiction between native and French precautions against the disease. Microbes and divinities, the natural and the supernatural, could both work in the world to cause human suffering and both needed to be addressed. The medical administration, for its part, continued to be baffled by this attitude, as the next section reveals.

## Magic and Politics

In 1912, a woman named Neang Tuch solicited the aid of the medical chief of Battambang for an unusual medical case. She had requested the French court hear her case of sorcery after the Khmer courts had failed to give her satisfaction. Her neighbor had cast a sickness spell on her, been convicted by the Khmer court, and ordered not to cast any more spells on her. When Tuch again found herself sick two months later, the Khmer court ruled that the neighbor had not caused the illness. Tuch thus sought redress with the higher, French court system, which chose not to hear her complaint. In her request to be heard by the French court, she described the process whereby a local monk extracted a large number of needles and other magically placed objects from her body; she also provided the names of several witnesses to this healing ceremony. Although the French courts seemed to consider the case ridiculous, the French doctor—out of curiosity perhaps—investigated her case, interviewing the woman, her witnesses, and the monk. The doctor ultimately declared that only superstition was at work. Tuch clearly did not recognize the "nonsecular" nature of magic and its exclusion from the French juridical system.[101] The French court may have viewed her efforts as ridiculous, but considering that the Khmer courts regularly ruled upon issues of magic, Tuch's appeal to a higher court was not illogical.[102]

The colonial government did not totally disregard all things magical. In several instances administrators sent out investigators to report on rumors of magic, many involving medicine. However, these investigations were motivated largely by political concerns rather than medical or scientific curiosity. Those with convincing magical powers had followers. Followers could mean social movements, and social movements could be dangerous. For example, two prominent charismatic millenarian leaders who led rebellions against the Khmer throne and the French colonial government in the previous half-century claimed to have healed themselves of cholera; this skill was integral to their perceived power. Leaders of other religious-cum-rebellious movements also claimed either the possession of instruments or the capacity to heal and protect followers from harm.[103] When these medical messiahs did not represent a political threat, the colonial government left them alone. On the other hand, when the administration saw any link to political activity, these outfits were quickly repressed. For instance, due to the continual uprisings of ethnic tribes in the Haut Chhlong region, the administration rapidly opened several investigations around Moi sorcerers selling lustral water in the late 1930s. Even though these were usually small-scale magical networks, the French jailed several of the "pretenders" who were leading these operations.[104] Contrast this to the case of a deformed young child referred to as *kru* (teacher) Hom. This young boy developed a large following in the Battambang area after he had healed his aunt of a chronic disease through a medicament revealed to him in a magical dream. Young Hom had quickly accrued a large following, a mass of wax,[105] and other tributary gifts, but showed no signs of any political inclinations. He also practiced in an area and at a time where there was little political agitation. After an investigation by the French doctor, the boy was left alone with his healing powers and his wax.[106]

## Better Health without Better Agreement

In adopting specific Western practices, Khmers showed no inclinations to drop their own beliefs—a "lack of change" that could be read as failure on the part French medicine. The ability to be all-inclusive of seemingly contradictory information would time and again baffle French administrators. French literalism, inflexibility, and excitability in the face of what Khmers considered as fundamental givens in the natural and supernatural world baffled the villagers, especially considering their own generally nonexclusive approach to French ideas. Khmer adoption of parts of French models of disease causation without abandoning their own traditional beliefs was to the French a mark of bad faith, illogic, or insincerity.

By the end of the 1930s, the country would witness a rise in reports of sorcery, both medical and otherwise—an indication that the indigenous population was not adopting the rational medical beliefs the medical administration was attempting to disseminate. Rhetoric generated for social medicine and changes made on paper often were not effectively enacted in Cambodia. Neither the message nor the measures penetrated beyond select enclaves. This was due to lack of money and AM resources, but also to the lack of educational infrastructure. However, the balance between acceptance and understanding of public health, so crucial to public health reforms in the industrialized West, was not the first concern in Cambodia. Further, the medical administration may have seen its own role as being the enlightenment or education of local populations, but in truth its efforts could still more accurately be labeled conversion. Attempts to change indigenous behaviors and ultimately indigenous understanding of health and disease prevention had to contend with preexisting and equally convincing epistemologies of health and disease.

However, public health was not a total failure. The indigenous population adopted some measures for improved public hygiene and eagerly adopted specific, widely distributed medications such as quinine and stovarsol. Also, a few towns received the benefit of modern sewage systems and improved potable water supplies. Both yaws and smallpox substantially decreased as epidemic diseases, although they were not eliminated.[107] Public health efforts ameliorated the living conditions of some, if not many. Further, the population of the country was increasing rapidly (although this was due to factors beyond the efforts of the health services). These perhaps were not the great successes hoped for by the administration, but they were to be expected. The system was planned and legislated, but it did not meaningfully operate beyond urban centers. In many ways, just as degeneration in the Cambodian countryside was an imagined construct, so too was the French public health system.

FIVE

# Prostitutes and Mothers

In most of our story thus far, more than half of the population, namely women, have been largely absent. This absence is in part an artifact of the historical record; French colonial health care providers were predominantly male, as were the patients. Women in Cambodia were generally unenthusiastic about French colonial medicine. The statistical records reveal that the vast majority of indigenous consultations and an even larger percentage of hospitalizations were men (figures 5.1–5.4). In medical districts with relatively good female attendance, women still accounted for less than 30 percent of consultations and rarely more than 20 percent of hospitalizations. In many isolated districts, their absence was even more pronounced. Children were rarer still. Even as the number of consultations and hospitalizations grew during the colonial period (both due to the structural growth of the AM and the growing population of the country), the ratio of women and children to men remained low.[1]

The state's focus on women's health fluctuated during the colonial period, growing in the post–World War I decade and gaining momentum in the 1930s. Until the 1920s, however, French doctors for the most part were untroubled by this disparity. When access to Cambodian women was deemed necessary, the medical services found this segment of the population elusive. Colonial medical interest in women and children and the methods used to intervene in female health bear on wider historiographical arguments in gender studies and the history of medicine. The reasons why indigenous women resisted these efforts also give us an opportunity to examine specific scholarly claims about the nature of gender and politics in Southeast Asia. This chapter makes three claims in dialogue with this body of scholarship.

First, concerns with male and population health ultimately motivated colonial efforts to improve female health, as a desire to ameliorate venereal disease among men and obtain access to reproductive capabilities underwrote these efforts. In other words, healthy women were necessary to a healthy military and a strong economy. Just as Ana Klaus has demonstrated in the context of France and America, we find in the colonial context that changing state populationist ideologies were attempting to redefine reproduction and childrearing.[2] However, ideologies alone did not determine the nature of this involvement. As one scholar observed, "Although the mother might appear to be the central subject of maternity, she is often evacuated

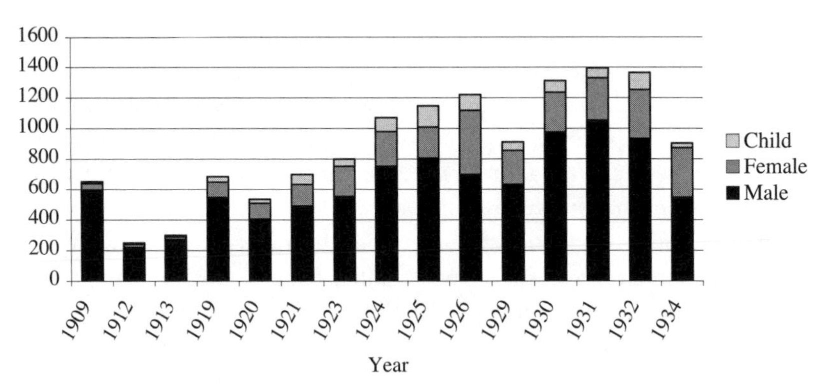


Figure 5.1. Annual Battambang hospitalizations, by gender and age.

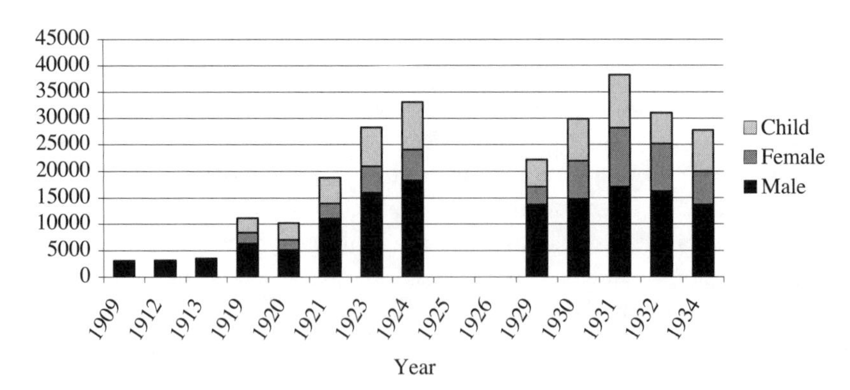


Figure 5.2. Annual AM consultations in Battambang, by gender and age.

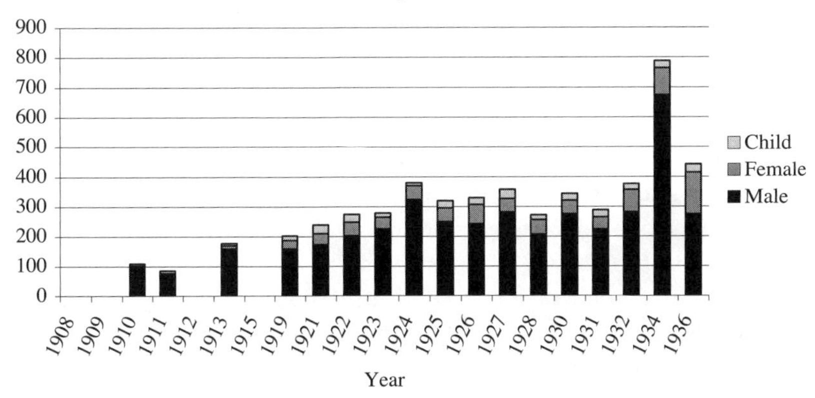


Figure 5.3. Annual Takeo hospitalizations, by gender and age.

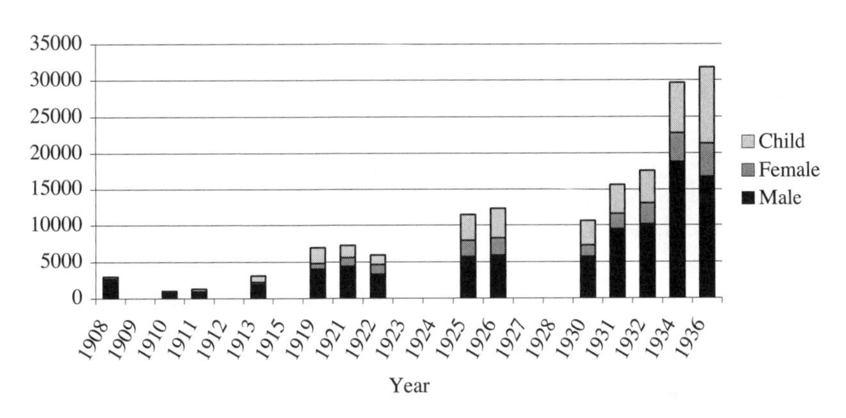


Figure 5.4. Annual AM consultations in Takeo, by gender and age.

from this position by a discursive focus on the child."[3] Despite women's peripheral status in the colonial state's agenda, their participation was central to how these programs on childbirth and welfare developed. While political constituency shaped the contours of medical intervention in France, public constituency at the very least influenced its tone in the colonies.

Second, the colonial engagement in women's health in Cambodia was closely tied to maternalist movements in the West. While European women imported the maternalist movement to their new colonial homes, its manifestation in the colony would reflect many of the contradictory ideals indigenous women faced in new public and private roles. This colonially-inflected maternalism was unique. Scholarship on the maternalism movement has observed that in Europe during this same period, "maternalism always operated on two levels: it extolled the private virtues of domesticity while simultaneously legitimating women's public relationships to politics and the state, to community, workplace, and marketplace."[4] European women played an active role in constructing this private and public status, wresting the discourse away from the need to correct maternal deficiency and towards the support of maternal nurturance. The role of indigenous women in Cambodia, by contrast, is harder to characterize. French "mentors" circumscribed their roles; their own cultures played only a small part in defining these new subject positions. While medical interventions and practices evolved in negotiation with women's participation or lack thereof, do the concepts of feminism and maternalism that underlie these practices follow a similar track? How does the history of maternal and perinatal medical interventions relate to the history of the discursive (re)constitution of "female" and "maternal" in the colony? An analysis of both European and indigenous women's reactions to medical interventions on their bodies must thus make a distinction between Western feminism and its manifestations in the colony. I hesitate here to call this "Third World feminism" or "indigenous feminism," as some scholars have done.[5] While some of the discourse and practices of the maternalist movement are apparent in Cambodia, it is unclear if feminism or a feminist impulse takes root among Cambodian women. On the other hand, is the political and social equality in existing gender relations such that a discourse on female empowerment is unnecessary? In other words, are Cambodian women already empowered and equal, or more equal, to men than in the Western context? This brings us to the third and perhaps most significant observation about the nature of women's medical intervention in colonial Cambodia.

Efforts to draw women to Western medicine and to train women as its providers, along with the ultimate failure of these attempts, expose the contradictions between the "modern" and "traditional" role of women in Cambodia. As Barbara Andaya comments of Southeast Asian studies, "in the current discourse, 'traditional roles' carry with them implicit messages of gender equality, economic independence, etc., and invoke a kind of golden age when women were different from men 'but in no way inferior.'"[6] This chapter questions this trope of a "golden era" of female gender equality in precolonial Southeast Asia that has been weakened by the intrusion of modernity and capitalism.[7] As we write women back into colonial history, we inevitably run into the problem of sources. Colonialism is often portrayed as a matter of "conquest, pacification, domination and exploitation, deeds carried out mostly, if not solely, by male actors."[8] While men may commonly serve as the actors in the histories we write, women are often reduced to symbols. Thus, the inclusion of women in colonial history often means capturing how women *are experienced* during a historical moment (by men), rather than how women themselves are experiencing the moment. Indigenous women occur everywhere in the historical record as symbols of social boundaries, of home, of danger, and of pollution—but rarely as agents.[9] Scholarship on colonialism and gender has established that European imperialists were "constantly at work to consolidate and secure an authority predicated on whiteness, maleness, and European-ness."[10] European imperialists discursively racialized and feminized the colonial subject to bolster European rule. In considering this phenomenon, historians who focus on indigenous Southeast Asian women and colonialism can fall into a conceptual trap. That is, to speak of the changing role of women in relation to the sexual politics of empire, we may be tempted to posit a less domestic, more masculine woman who can be "feminized" by the colonial endeavor along with her Asian male peers.[11] However, a study of medical intervention in the colony reveals that assumptions of a sexually egalitarian precolonial Southeast Asian society must be approached with caution. A history of medical intervention undermines this formulation. Efforts to draw women to the medical service as both providers and patients failed in large part because of significant inequalities in existing gender roles. Even as the French colonists projected a discursive ideal of woman on the colonial subject, the Cambodian woman acted with little regard to these projections.

Responses to the practices engendered by this discourse are a somewhat different story. Although the colonial government was attempting to transform masculinity and femininity, it on some level misread what it was attempting to change. Women, as they fell under the purview of Western medicine, shaped their own involvement in deliberate and unexpected ways. Precolonial gender identity was not predicated on social equality; thus, Western medicine could be both liberatory and oppressive.

## Protecting the Soldiers

The one major exception to the medical administration's neglect of women's health before World War I was in the control of prostitutes. The military monitoring of prostitutes was the first and longest running French colonial program dealing with the health of women—if only a very particular sector of women. Maurel, the doctor who credited himself as creator of the Cambodian medical service, instituted a service for prostitutes concomitant with the army medical service in 1885.[12] This effort was driven by a concern with the health of French troops, since venereal disease incapacitated up to a quarter of the troops at certain times.[13] Created to ensure the health of a key male population sector, this program highlights again the secondary nature of concern with women's health. While one historian has traced the lineage in Cambodia of sexual service from a concubine system of indigenous women with male foreigners to a system of contractual payment for short-term sexual service in the last quarter of the nineteenth century, these are actually two separate phenomena which involved different groups of women.[14] Long-term concubinage derived more closely from Khmer social structures and gender relations, while prostitution was a money-driven contracting of a brief sexual service. The women involved in the latter activity, who were often sex-trafficked, were foreign colonial imports from neighboring Vietnam as well as from more distant Asian countries.

The surveillance of these latter *filles publiques* quickly became standard to the medical service. The method of surveillance in many ways mirrored the regulation of prostitution in France. However, if prostitution in France became less criminalized and more medicalized in the twentieth century,[15] in Cambodia it became both more criminalized *and* more medicalized. Just as in France, prostitution itself was not illegal. Soldiers were urged to visit registered prostitutes in monitored *maisons de tolérance*, and were strongly encouraged to avoid unregistered prostitutes, *femmes racoleuses*.[16] Through the first half of the twentieth century the relatively lax regulation of prostitution became more rigorous, although it would remain legal throughout the colonial period. For example, late-nineteenth-century weekly visits by registered prostitutes at an open dispensary in the Mixed Hospital were gradually replaced with forced detention of the sick at a dispensary—the Dispensaire des Filles Publiques—attached to the Protectorate Prison, built in 1912.[17] In addition, the *maisons de tolerance* instituted by Maurel at the end of the nineteenth century were gradually phased out.[18] Before the 1920s, prostitutes acquiesced to occasional medical detention at the dispensary despite its proximity to the Protectorate Prison. When the new Lannelongue Dispensary for prostitutes opened in 1925, ostensibly to improve patient care, it provoked intense negative reactions (figure 5.5). A riot—the first large-scale protest by these women in the history of prostitution's regulation in the country—ensued shortly after the "improved" facility opened. It began when six "patients" attempted escape and were forcibly returned to their rooms.[19] The new dispensary, although better built than the older facility, operated in a more bureaucratic fashion. Significantly, it also had bars enclosing the entire first floor verandah. The patients-cum-inmates wore uniforms and followed

Figure 5.5. Construction of Lannelongue Dispensary, 1924. The bars along the verandah are partially installed. Courtesy of the National Archives of Cambodia.

a regimented routine. In the health director's opinion, these changes made Lannelongue Dispensary "seem" more prison-like to the patients, leading in its first year of operation to the riot as well as to several other dangerous escape attempts, protests, and collective "neurasthenia."[20]

While early government interest focused predominantly on prostitutes, the private colonial sector showed some interest in women and children before the interwar period. However, this interest came from other areas of the French colonial society, namely the Catholic Church and the wives of colonial administrators. Just as the desire to improve the health of male troops motivated the medical regulation of prostitutes, the concern with child and population health drove medical intervention among indigenous females. Almost all medical efforts focusing on the female sex involved reproductive health. The AM created *no* health programs targeted at unmarried women (except prostitutes), the childless, or the aged. While France as an imperial power may have been concerned more with cultural than economic production in its colonies,[21] the nature of French medical intervention among Khmer women was likely entwined with both types of production. In other words, the medical interest in children correlated with the economic focus of the colonial endeavor, while the French medical imaginings of femininity and maternity reflected, or at least refracted, French cultural ideals of the mother.

## The Catholic Necropolis

The Catholic Church was arguably the first Western medical service offered in Cambodia (Maurel's ambulance being the first *government* medical service). The order of the Sisters of Providence (Soeurs de la Providence) established a crèche, orphanage, and hospice in Phnom Penh in 1881 and a similar hospice in Battambang in 1905. In stark comparison to the military hospital, these religious establishments had a relatively equal ratio of male to female patients.[22] After the AM formed in 1907, the order continued to operate its crèche, but the colonial administration forced the Sisters to agree to AM medical monitoring in return for a government subvention.[23] Church cooperation with the medical services was half-hearted at best, and its interest in the *physical* well-being of its wards was suspected of being secondary to its interest in spiritual care. Further, the church focused its efforts largely on the Vietnamese population, which made up approximately 90 percent of the congregation. On several occasions the health director accused the Sisters of deliberately turning away Khmer women in favor of Vietnamese.[24] The predominantly Vietnamese congregation was not entirely due to church selectivity; the Khmer population seemed less than eager to take advantage of the services offered by the order. Quite early in the colonial period, the Catholic Church had focused its missionary efforts on ethnic Vietnamese, who had proven to been more open to Catholic conversion than the Theravada Buddhist populations of Southeast Asia.[25] In fact, the Sisters designated to serve in Cambodia were first trained in Culaogien,[26] Cochinchina, where they learned not Khmer but Vietnamese before reporting to Phnom Penh (figure 5.6).[27] Neglect by the church was probably a boon—physically, if not spiritually—to Khmers. The AM was to repeatedly accuse the Sisters of gross medical negligence. The order also continually refused to comply with AM reporting standards.[28] For example, in 1919 the 309 deaths at the Providence crèche represented more than one-third of the child mortality of the entire city of Phnom Penh.[29] The municipal doctor would note that these deaths were attributed to "fantastic diagnoses . . . which are provided by a Sister devoid of any medical knowledge and ignorant of the very words she uses." He would continue: "Some of these children remain several days in the establishment and have no care other than the attention of this sister who, as is fitting for her position, is preoccupied more with saving souls, with a little angel to send to heaven than with judiciously distributing medicines or applying a therapy that might save the body."[30]

Figure 5.6. Soeurs de la Providence. Culaogien, circa 1890.
Courtesy of the National Archives of Cambodia.

Observing that the church had never once during the year called for medical assistance while it continued to fill a field behind the crèche with thousands of cadavers of children who died without care," the doctor decried such medical assistance, stating that the health services "with all of our force . . . claim our indignation in the face of these facts."[31]

The Sisters remained committed to their program of "making angels" despite criticism from the French medical establishment.[32] In 1926, Hygiene Director Louis Simon was again to express outrage that the crèche had registered a 100-percent mortality rate among its infants for several years. That year, he forced the Sisters to allow regular monitors from a local women's charity; however, the visits only decreased infant mortality by one percent.[33] The Health Service was relatively consistent in its lack of enthusiasm for the Sisters. As late as 1938, another health director would again ponder the government's financial assistance to an organization with a nearly 100-percent mortality rate among its infant wards.[34]

The generally negative stance of the medical service towards the Sisters was likely a mixture of professional outrage tinged with anticlericalism.[35] The RSC, on the other hand, was steady in its financial support to the order for its charitable works throughout the colonial period, although this may not have reflected steady political support. The damning charges against the Sisters were in part tempered by the state of the children brought to the crèche; from all accounts, most of them were in terrible physical condition due to sickness or neglect. The Sisters served an additional and perhaps more useful social purpose by providing care and shelter for aged and chronically infirm Vietnamese; however, they had little positive impact on infants' or women's health. The second medical charity to which the RSC showed steady financial support was a semiprivate women's organization to which we will now turn.

## Charitable Works and the Colonial Housewife

At the turn of the twentieth century, colonial medical interest in the Khmer female population came largely from French women: homemakers and wives of colonial administrators. Whereas the Catholic Church's involvement in medical care was more a door to spiritual well-being, the interest from these French women could be characterized as a genuine interest in corporeal health. However, it would be more precise to call it an interest not in women's health so much as population health.

On December 9, 1906, several colonial administrators and their wives formed the Society for the Protection of Indigenous Births (Société de Protection de la Natalité Indigène, or SPNI), under the honorary presidency of King Sisowath and the patronage of RSC Luce, Phnom Penh Mayor Paul Collard, and the *résident* of Kandal, Gabriel Jeannerat. The society's purpose, as outlined in its bylaws, was to encourage women to give birth in the free maternity, to provide pregnant women and newborns with physical care, to provide mothers and their children with material aid and clothing on departure from the maternity, and to train indigenous midwives to give care according to the prescriptions of French medicine.[36] A few years later, the group changed its name to the Society for Maternal and Infant Protection in Cambodia (Société de Protection Maternelle et Infantile au Cambodge, or SPMI) in order to reflect an interest in the health of indigenous mothers as well as their young children. The name change may also have been motivated by a desire of the society to distance itself from a similarly named organization whose focus was also on children in Cambodia: the Society for the Protection of Infancy in Cambodia (Société de Protection de l'Enfance au Cambodge, or SPE). Unlike the SPMI, the SPE concerned itself solely with the mixed offspring of French fathers and indigenous mothers. Originating in Saigon in 1894 as the Society for the Protection and Education of Young French Métis in Cochinchina and Cambodia (Société de Protection et d'Education des Jeunes Métis Français de la Cochinchine et du Cambodge), it formed a separate branch in Phnom Penh in 1904.[37] Similar organizations also existed in other parts of Indochina. These societies openly encouraged French settlers to donate support and money to aid these mixed children in order to prevent them from disgracing the French race.[38]

In 1907, due to the low membership and lack of public interest, Phnom Penh's Society for the Protection and Education of Young Métis French renamed itself by dropping the explicit reference to mixed French/Indochinese children, hoping that the change would give "strength and extension to our charitable works."[39] This ploy seemed to work, as the SPE gained more local interest over the next several years. However, the change was in name only. The perceived problem with métis children would remain the SPE's focus through the colonial period; its main work would be to find and raise them.[40] Its interest in their mothers was minimal.[41] However, the work of the SPE would come together with the work of the SPMI when, in the 1920s and 1930s, the SPE's female wards would be funneled into the midwife training programs of the SPMI-created maternity. The tensions created by the racially distinct groupings of the two organizations would become inverted in the training of SPE women as SPMI midwives, as we will see at the end of this chapter.

The SPNI, before its name change to SPMI, quickly built a Western-style maternity, which opened on September 12, 1907, in Phnom Penh, staffed by an aged Khmer traditional midwife (*chhmap*) and monitored by the French doctor.[42] Although the service was free, the society realized that incentives were necessary to attract indigenous clientele. Thus it allotted a subvention of one piastre to each mother who gave birth in the maternity, along with a gift of a layette for the newborn.[43] At the order of the RSC, the Khmer Council of Ministers distributed an announcement on the new maternity to local village chiefs. Written by the Khmer elite at the behest of French administrators, the notice boasted of the largesse of the French, who, having determined that the people of Cambodia were poor, had created a free service for pregnant women. This service would be performed in a new "public birthing house" (*munti somraal*) called a "*maternité*" (transliterated in Khmer). In contemporary usage, *munti somraal* is translatable as "maternity hospital," but this initial combination of *munti* (public office or bureau) with *somraal* (to give birth) must have had a radically new meaning at the time. Until this moment, birth had been a private ritual within the home, attended not by doctors but by the *chhmap* and family members. This announcement marked birth as an explicitly medical process requiring government involvement.

The birthing house, the announcement explained, had been built because the French administration was "good-hearted" and the charitable organization that sponsored it wanted to "earn the devotion of Khmer women" by ensuring that their newborn children would grow up "happy, strong, and healthy."[44] There were no men in the maternity to embarrass or watch the women, with the exception of a French man "who will care for the woman more than she would be cared for at home."[45] Interestingly, the Khmer announcement, unlike the French, discriminated in the amount given to the mother according to the sex of the child born. The birth of a boy warranted a five-riel[46] subvention (approximately the equivalent of one French piastre), a girl only "2 to 2.5." Although this difference did not exist in any of the French documentation, the colonial government approved the poster that announced it. This admittedly small occurrence hints that, at least among the elite, the much-vaunted equality between the sexes in precolonial Southeast Asia had its limits.[47] This initial advertising was not greatly effective. In its first year of operation, the maternity had on average less than one birth a day.[48]

At the beginning of the twentieth century almost all human births around the globe occurred in the home, including those in France and Cambodia. The number of home births began to decline in the first half of the twentieth century in the industrialized West, but even as late as World War II, most births in many Western nations were still in the home.[49] While the global shift towards hospital birth was occurring precisely at this moment, this transition was much slower in the colonies. The story of the SPMI maternity reveals some of the dynamics of why rates diverged at this historical moment.

The SPMI maternity was renamed the Ernest Roume Maternity on June 30, 1919, when the medical services took control and annexed it to the Mixed Hospital.[50] This annexation was unsurprising, considering that the charity had never intended to become a medical organization. Knitting baby hats (a commonly mentioned, though puzzling, accessory in the heat of Cambodia) and distributing powdered milk were tasks entirely different from performing births or caring for medical complications that could arise with pregnancy. The SPMI continued to provide advice and baby supplies from its headquarters in the mayor's office into the 1930s, while endorsing births at Roume.[51] However, it continually refused the health services' offer of free consultation to its clients. This attitude, in the opinion of the health director, indicated that the SPMI's work was only a pretext for a self-congratulatory women's social group.[52] Most likely, the refusal of assistance stemmed from a desire both to maintain group autonomy and to encourage attendance by providing a women-only environment. The organization was successful enough to create a second branch in the seaside town of Kampot in 1925.[53] It also continued charitable works in other guises, including several explicitly eugenic programs that became fashionable in the 1930s.

Since the AM doctor had previously monitored the maternity when it was under SPMI control, Roume continued to operate in a similar manner after being transferred to the health services. Its clientele steadily increased as segments of the indigenous population became more familiar with its existence and purpose. The population of Phnom Penh nearly doubled from fifty thousand in 1909 to slightly less than one hundred thousand in 1939.[54] Maternity births increased tenfold in the same period, as figure 5.7 reveals. Some small variations occurred. The 1933 drop in births was due to the disruption of medical service with the temporary contraction of the Cambodian AM into the Cochinchina service for the period of January 5, 1933, to June 9, 1934.[55] The dip in 1937, on the other hand, corresponded to the suppression of monetary aid to new mothers at Roume. In that year, the health director abolished the one-piastre payment for births, arguing that this small aid was a constant cause of argument and disorder, with rumors of theft by personnel and accusations by indigenous mothers who did not receive their money. He reasoned that Roume was by this point

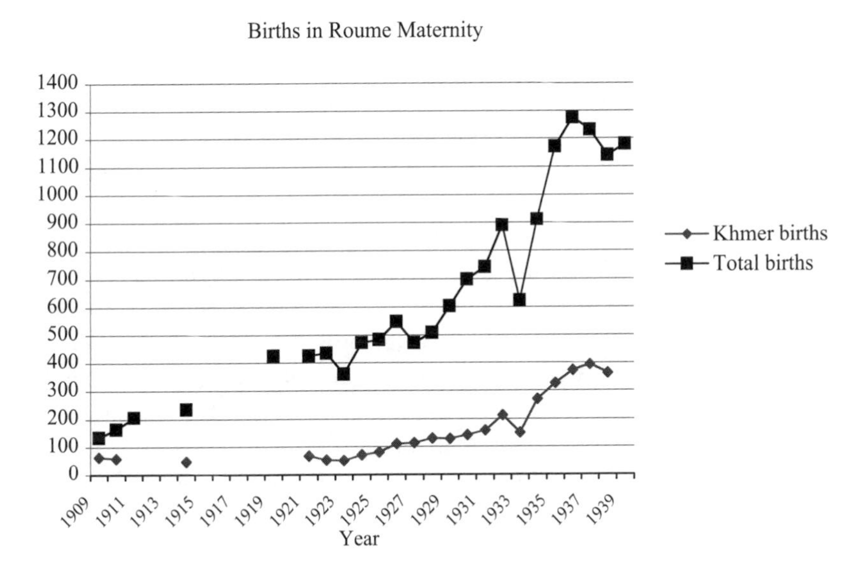


Figure 5.7. Comparison of SPMI and Roume Maternity births.

well enough advertised to cease such charity. He would comment almost lackadaisically, "If we find the number of hospital births diminished by this necessary measure, we will not be unduly affected either from a moral point of view or from a service point of view."[56]

The use of the maternity differed remarkably among ethnic groups. The Vietnamese population made up the lion's share of the tenfold increase in maternity births between 1909 and 1939, with the Chinese and Sino-Khmer women providing another large segment. Khmer women, in contrast, remained unenthusiastic (figure 5.7). The fragmentary statistical records by ethnicity for consultations and hospitalizations reveal the same patterns in general AM usage. Thus, not only were women as a whole greatly underrepresented within the medical statistics, but Khmer women were themselves a minority within that small group. Some of this unevenness can be attributed to the large number of Vietnamese men in the French bureaucratic structure. Hesitation to resort to French services was likely lower among those who were already familiar with the colonial government. The families of these men had more contact with all things French. Further, the predominantly Vietnamese auxiliary medical staff may also have played a part. Equally important, this disparity was also due to the cultural role of Khmer women versus other ethnic groups within the colonial social structure, a phenomenon to which we will return.

## Depopulation and Underpopulation

The medical administration's concern with the health of women grew in the 1920s and 1930s, following the growth of the ideals of "social medicine" that came into vogue in the metropole. The national mood in interwar France was bleak. Societal concerns with underpopulation and national decline throughout Europe and North America led to an interest in women's reproductive health and eugenics.[57] Historian William Schneider has convincingly argued for the strong tie between natalist and eugenic movements in metropolitan France.[58] Unlike eugenics in the United States, Germany, and England, which developed in close relation to anthropology and biology, the French variety was unique in its origin in the new science of puericulture, or infant care. Puericulture itself, not coincidentally, had begun in nineteenth-century France, and many of the 1912 founders of the French Eugenics Society were prominent doctors of puericulture. In the French context, eugenics and the perceived problem of depopulation underwrote the goals of social hygiene. The renowned natalist Jean-Louis Bréton was chosen to head the first cabinet-level health ministry in France. Further, the Social

Hygiene Office, founded in 1924, originated from the work of the Rockefeller Foundation's French tuberculosis program and the French Eugenics Society.[59] The connections between social hygiene, women's health, and eugenics would reverberate in the medical and social policies of colonial Indochina. Colonial administrators, in applying these policies in the colony, on some level conceived of their subject populations as French populations. However, the most important of these populations were clearly the economically productive segments.

Once the low number of women using colonial medical services was deemed a problem, AM medical reports began to offer reasons. The medical services failed, however, to recognize what was likely the greatest explanation for the gender disparity. Much of what we have discussed in previous chapters on native health has dealt largely with men's health, and only with a specific sector of men. Because we have been speaking of the interaction between Khmers and Western-trained doctors, we have been dealing largely with the public sphere. Many Khmers were not entirely voluntary patients. They came mainly from the military, government offices, public work sites, administrative schools, prisons, and plantations. These were all male-dominated public spaces. Thus, Western medicine accessed indigenous bodies through the enclaves of colonialism. Such enclaves were predominantly male, and disproportionately ethnic Vietnamese. Like the zenanas of India or the harems of Morocco, the "female" spaces of Cambodia were cloaked from the imperial gaze.[60]

With the exception of marketplace activity, the social role of Khmer women was largely confined to the domestic sphere. Their interaction with the colonial government was rarely ever direct, and largely incidental. Young women particularly were shielded from interaction with most men. Foreign men were doubtless more taboo. French colonial scholar Etienne Aymonier, in compiling a survey of Khmer customs, found that obtaining permission to simply observe young women from a distance was difficult, and to speak with them impossible.[61] Except for their relatively high attendance at vaccination *séances* for their children, Khmer women had little contact with the French doctor. One notable exception would be the repressive measures in place during epidemics, but even then the doctor usually did not gain access to the female body until her death.[62] The attitude of mothers carried over to their children. Mothers used domestic treatments on their children, or opted for one of the variety of indigenous medical practitioners available. Pregnancy and childbirth were the province of the traditional midwife, the *chhmap*. A survey of traditional birth rites reveals the vast gulf between the meanings and methods of Western and indigenous health practices.

Medicine, and particularly childbirth, was centered in the home and steeped in ritual. Women gave birth at home, aided by a *chhmap*. Pregnancy and its aftermath were considered states of heightened danger. Before and during pregnancy, Khmer women believed themselves particularly vulnerable to evil spirits, *priey kraalaa pleung*.[63] Therefore, before and after birth they underwent various rituals to insure their own health and that of their new children.[64] Protections against spirits could include incantations, ritual application of lustral water on the pregnant mother, or ritual placement of protective items around the home. A parturient woman invited bad luck if she inquired about other pregnant women, particularly if the other women had had health problems or accidents. More generally, the pregnant woman needed to stay away from any site of sickness or recent death.[65] This clearly would have been a prohibition that the most flexible hospital could not accommodate. Labor itself was the time most fraught with spiritual and physical danger; the birth space needed to be a "closed world,"[66] both physically and magically protected from outside intrusion. Men were generally excluded from the birth area.[67] The woman was normally accompanied during labor by one or two very close relatives, a midwife, and possibly a healer. Birth depleted the mother's physical energy, or *sasae* (literally, "threads" or "muscles"), making them raw or new. To restore these threads, a new mother had to rest immobile on a bamboo bed built over a small brick stove, "roasting" for several days to several weeks.[68] When the woman's threads were renewed, the mother, family members, and in some instances the attending midwife ceremonially extinguished the roasting fire.[69] As aptly observed by Menaut, such rituals could not be performed within a modern hospital. However, the woman—the "patient"—clearly saw them as significant. Giving birth in a hospital was thus not simply a change of venue; it was a total transformation in the meaning and process of giving birth.

The French medical service did not comprehend the transformative aspects of hospital birth. Rather, it tackled what were secondary causes of the unpopularity of Western medicine. Indigenous women were poor, so services would be free, and a gift or bribe would be offered. Women were afraid of men, so women would be hired to treat women. Young women were "timid," so they were coaxed with promises of special care and polite service. These were superficial changes to address superficial cultural differences. Certain other efforts to popularize Western medicine, as we will see below, were removed even further from local context. Policies stemming from anxiety over depopulation and degenerating human stock in the metropole were dropped into colonial Cambodia.

## Wooing Female Clientele

In 1928, the Roume attending doctor observed that the majority of his Khmer clients ended up at the maternity by chance or in a state of dire need. He believed that four main factors prevented their voluntary attendance. The first was the "extreme modesty" of Khmer women, even among the educated and ruling classes. Thus, he had found greater success by allowing the midwife to perform initial consultations alone and call him when needed. Also, the growing European, Vietnamese, and Chinese populations were pushing the Cambodian population further and further from the center of the town, where the maternity was situated. This distance increased both the cost and the difficulty of getting to Roume. Further, he believed that the maternity's enormous size frightened the "timid" Khmer women. He finally offered a biological reason for the absence of Khmer women: they, unlike Chinese and Vietnamese women, were physically strong, lived in the open air, worked for a living, and generally had more fortitude and health to bear children without medical intervention.[70]

Much of this reasoning is echoed in other reports. The Prey Veng doctor would decry the rudeness of his indigenous personnel—particularly his midwives, whose behavior discouraged women from coming to the hospital. Like the Roume physician, he believed that this was not a serious problem, since Khmer women did not have the postnatal problems of other "races."[71] Patients also expressed strong dislike of regulations and enclosure in the medical establishment.[72] Certainly, many of the women who gave birth at Roume were eager to leave; in 1924 the medical services investigated midwives for taking bribes from Khmer patients who were desperate for early release.[73] This also indicates that patients at Roume were often there involuntarily. In 1926 the Kompong Thom doctor ascribed the lack of female patients to the lack of female personnel, since women were highly suspicious of revealing themselves to men.[74] In 1927 women from town refused to go near the Ang Duong dispensary in the center of Phnom Penh because, as built, it encouraged peeping toms.[75] When in 1936 the newspaper *Viêt-Nam* ran an unflattering exposé on the Phnom Penh Mixed Hospital, Health Director Simon would dismiss most of the article's validity in a letter to the RSC. However, he would write, "one thing the paper is exactly correct on is that there are NO female attendants . . . in the waiting room, and all the rooms reserved for women are filled with men."[76] He would also note that the country had no female doctors, so that efforts to accommodate the desire of women to be treated by women were limited. In the same year a Khmer *médecin* would comment, "Our women, little evolved, are still excessively prudish. There are those who prefer to die rather than allow themselves to be examined by a man." He would continue, with disgust, "She only brings her infant for a medical visit in the case of grave gastro-intestinal maladies or troubles, and then only after having exhausted beforehand all Sino-Annamite and Cambodian therapeutics."[77] During epidemics, when doctors forcibly examined dead bodies, objections to these doctors' viewing and handling of female bodies were much more common than objections to their general treatment of the dead.[78] Disrespectful treatment of female corpses would create more uproar than forcible evacuation of the living to treatment centers. During a 1908 plague epidemic, an angry crowd accused French doctors of taking the victims "to see the genitalia of the women, on whom they then wanted to have experiments [experiences], attempt fantastic operations."[79] As a response to the "violent repugnance" with which Khmers reacted to French doctors seeing the corpses of their wives and daughters naked, a French administrator suggested that a family member wrap a *krama* (thin cloth scarf) around and between the legs of deceased women (a substitute for Western-style undergarments) before the doctor removed her sarong to examine her.[80]

Along with the sexually-charged suspicion of men, race was a major issue. Again, colonial observers also blamed racial animosities for Khmer avoidance of both Europeans and Vietnamese. After the Vietnamese midwife was fired from Roume for grave negligence in 1920, the attending doctor, Hervier, conveyed an urgent message to the GGI that the "repugnance of the indigenous female for European care" would create serious problems if another native midwife were not found.[81] Most patients at Roume were Vietnamese. Thus it would seem that in the French doctor's perceptions, Vietnamese women, like Khmer women, preferred to be treated by their own. Again, this may have been a factor behind the predominantly Vietnamese births at Roume. Observations in later years would also highlight racial animosities. For instance, the Svay Rieng medical chief believed that an Annamite midwife was "able enough," but that her action was limited to the Chinese and Annamite populations since Khmers would not come to her.[82] A 1928 Phnom Penh report would make the identical observation, adding that the French midwife was the object of the "same ostracism."[83] Dr. Menaut, recognizing the "antipathy between the Annamite and Cambodian races," attempted to take precautions at the maternity so that one group did not treat the other, but he found that they were not effective. He would resignedly observe, like his predecessor of previous years, that the only Khmer women to come to Roume were "women abandoned by the father of their child, unfortunately without a family, women surprised by initial birthing pains during the course of a voyage and not knowing where to take refuge to protect their bodies, and crossing with fear the threshold of this fearsome building. . . . Many reasons explain the Cambodian's apprehension, first a question of modesty and timidity, but also a question of respect for ancient customs, to which the race remains strongly attached."[84] These rituals, he observed, simply could not be done in the orderly environment of a hospital. Indigenous *médecin* Yinn Vann agreed. He believed that by accommodating traditional practices, "we would possibly have more clientele giving birth at the hospital . . . ." He hedged, however, saying that openness to superstitious customs, "too contrary to the rules of sterility in any case to tolerate . . . , [would] risk infections for which we would then be responsible."[85] In 1934, the health services in conjunction with the SPMI began studying ways of enlisting monks to encourage women to come to the maternity. The study found that the largest deterrent was the conflict between traditional birthing rites and the methods of the hospital.[86] Although the centrality of ritual was one of the most obvious reasons for female avoidance, the French administration did not address the issue systematically until the late 1930s. Initially, it attempted to advertise and promote the Western maternity.

## Better Babies, and More of Them

Having identified numerous reasons for women's avoidance, the AM employed several tactics to bring them under medical purview. However, these tactics all focused not so much on obtaining access to the woman, but on obtaining access to her reproductive capacity. And again, this was connected to the growth of the hygiene movement and the concern with population health across the globe. In 1921 the Cambodian annual medical reports began explicitly mentioning social hygiene as a goal of the AM, which would approach it in three ways: medical works, public education, and the development of societies for infant protection.[87] The inclusion of infant protection as a major goal reveals the importance of population thinking among policy makers, and its ties to group rather than individual concepts of health. As traced in the previous chapter, policy makers discussed and legislated much on public health behaviors in the 1920s and 1930s. Before World War I, the societies for infant protection had been largely private initiatives in Cambodia.[88] Social hygiene and infant protection were intimately related to eugenics; the AM maternities in Cambodia were explicitly referred to as eugenic domains.[89] The French colonial administration was not only concerned with better babies; it also wanted more babies. It continually perceived Cambodia as an underpopulated region, particularly in comparison with neighboring, densely populated Cochinchina. Despite the fact that Cambodia's actual population was growing tremendously during the colonial era, the colonial administration assumed that the "underpopulation" of Cambodia was directly related to low survivorship of Khmer infants.[90] The SPMI formed in part to combat this high infant mortality.[91] This perceived underpopulation also underwrote policies encouraging immigration from neighboring, "overpopulated" Cochinchina.

Medical monitoring by the doctor was the clear solution to the problem of infant mortality. Other solutions included improving the stock of infants born (i.e., selecting healthier babies), and educating indigenous women on how to make their babies stronger. French colonial interest in eugenics, maternities, and population health came together most obviously and interestingly in the Roume baby contests of the 1930s.[92] These contests also served as a means of publicizing the "modern" way of producing and caring for children.

The SPMI, in conjunction with the Roume Maternity and the colonial administration, began holding annual Concours des Bébés in the 1920s, which continued through the 1930s. Naturally, only babies born at Roume were eligible. The SPMI selected a few hundred babies each year to compete (figure 5.8). The judges were members of the SPMI and local notables. For example, the 1935 competition had three hundred babies who were judged by three French women of the SPMI, Khmer Princess Suramith, and two doctors of the AM. The winning babies and their parents that year received three bottles of Nestlé milk, one box of flour, a bolt of cloth, a blanket, a towel, a bar of soap, and "a beautiful little bonnet knitted by charity."[93] The corporate nature of sponsorship is quite interesting; the award of bottled milk to the best child/mother pair was rich in symbolic irony. As one study of Theravada Buddhist populations observed, "The most potent and unassailable symbol of protectiveness and loving kindness was that of a woman nursing a baby at her breast."[94] Considering this cultural symbolism, only the neglectful mother would deny her breast milk to her baby. In encouraging supplementation with bottled milk feeding, the medical staff did not apprehend the negative significance of such a "prize."

The *Echo du Cambodge* reported on the event with some levity: "The task of the jury . . . was not without difficulties, however . . . after a laborious examination [the jury] awarded 36 chubbies [*poupons*], who untroubled by their glory sucked on delicious bars of Nestlés chocolate."[95] The entire

Figure 5.8. Winners of baby contest during Varenne's visit, 1925. Courtesy of the National Archives of Cambodia.

competition was sponsored by the Nestlé Corporation. Other colonial newspapers were not so enamored with these contests. *La Presse Indochinoise* would report: "This competition, reserved for infants born at Roume, also likely serves to control and collect information on the sanitary state of Cambodian infants."[96] The newspaper, in its usual caustic tone towards the health services and the colonial government, painted the government's self-congratulation over its work on infant health as entirely overblown. The article further suggested that the entire *Concours* be renamed the urban baby competition, since only maternity babies were eligible and no maternities existed in the countryside.[97]

*La Presse Indochinoise* exaggerated the lack of maternities in the countryside, but not by much. Most of the provincial medical centers may have had rooms where births were scheduled, but none except Battambang and Pursat had a dedicated maternity building.[98] In part for this reason, most outlying medical districts often had fewer than ten hospital births each year.[99] For instance, in 1921 Phnom Penh witnessed 424 births while the remaining fourteen medical districts registered a total of 43 births.[100] By 1934 the gap had narrowed, with 912 maternity births in the capital, compared to 290 in the countryside.[101] By 1939 the number of hospital births in the countryside had increased substantially, with 1179 at Roume and 840 in the rest of the country. This was, however, still a miniscule number for a rural population of nearly 2.9 million. Only one in every 1,500 births occurred in a French medical establishment.[102]

## Women to Draw in Women

The Roume baby contests were a popular annual Phnom Penh social event, but they were not terribly successful in luring new female patients to the hospitals. As the *Presse Indochinoise* articles so pointedly stated, the contests also did nothing for maternal or infant care outside of the capital. The overwhelming agreement among AM staff was that Khmer women were hesitant about French medicine, a hesitation exacerbated by a suspicion of male (French or Asian) and non-Khmer female doctors and nurses. Thus, the logical solution was to find and train Khmer women to treat Khmer women. Before the interwar period, the AM recruited female staff in conjunction with its recruitment of men, as part of its augmentation of the native sector of the health service. During the 1920s the RSC launched several medical training programs specifically targeted at women. In sum, through the colonial period, three institutions trained women for the Cambodian health service: the École Pratique de Médecine Indigène in Cholon (the Cholon School); the Hanoi École de Médecine; and the several training programs at the Phnom Penh Mixed Hospital. Although several Vietnamese women who trained at the École de Médecine eventually served in the Cambodian AM, no Cambodian woman was ever trained at the École.[103]

On August 25, 1903, the GGI created the École Pratique de Médecine Indigène in Cholon, Cochinchina. This school's purpose was to train indigenous medical support staff (*infirmiers* and *infirmières*)[104] for the AM. Initially it focused on trainees for the Cochinchina medical service. The female staffers in the Cambodian service were, at this time, trained on the job. However, the supervising doctors would note the difficulty of keeping "Annamite" nurses through the training process. Many resigned or were fired, and their replacements disappeared a few days after being hired.[105] To formalize the training of Cambodian female staff, in 1915 the governor of Cochinchina reserved eight spots at the Cholon School for female students from Cambodia. Candidates had to be between eighteen and twenty-four years old, with an attestation of good manners, a medical health certificate, and a birth certificate.[106] Due to a lack of candidates, the allocation was reduced to five the following year. Between one and five students matriculated each year (except in 1917, when eight candidates were sent) and few completed the program.[107] In 1919, the Cholon School expanded with a program specifically for midwife training at the local maternity.[108] Applicants continued to be rare, and they rarely graduated. By 1924, the Cambodian AM employed only eight midwives: two trained in Hanoi and six in Cholon, all of them Vietnamese.[109] Khmer students were rare because Khmer "families do not readily consent to sending their young daughters to Tonkin or Cochinchina for their studies."[110] As it had done with male staff, the AM recruited ten royal princesses in 1924 as assistant birth attendants at Roume in hopes of raising the job's prestige.[111] As had happened with the male staff, this strategy was not successful.

Having taken over the Roume Maternity in 1919, the health services began planning a Cambodian school to train midwives, in the hope of overcoming the deterrent of foreign study. Centered at the Roume Maternity, the École Pratique des Sage-Femmes opened on September 17, 1924.[112] Candidates were required to read, write, and speak French; hold a primary school certificate; and be between eighteen and twenty-five years of age. Due to the lack of candidates, however, the health director immediately had to make exceptions in these entry requirements. This did not bode well for the quality and quantity of students: the first class matriculated five, with three more added in the spring semester (figure 5.9).[113] One student was admitted at thirty years of age, and another had no primary school certificate.[114] The employment results of these first three matriculated classes show a steady record of failure.[115] Even though students who left during the course of the program were required to repay ten piastres for each month of training completed, this did not deter dropouts.[116] Only six of the original eighteen students graduated. Of these six, only two had a record of extended employment within the medical service.

To further complicate issues, the new school exacerbated internal divisions among the existing female staff of the AM. Because the training at Roume was deemed more stringent than that of the Cholon school, the new Roume diploma was classified higher. The school's graduates entered the AM as midwives (*sage-femmes*) rather than as nurses (*infirmières*). The midwives' salaries began at 360 piastres per annum and topped off at 996 piastres, whereas *infirmières* began at 216 piastres and only reached 450 piastres per annum. Graduates of Cholon, although they often performed the same tasks as midwives, were classified as *infirmières* and received a considerably lower salary. To alleviate complaints, the health director allowed the eight preexisting *infirmières* to take an exam to switch classifications.[117] All who attempted to convert cadres failed the exam.[118]

Admissions did not improve in future classes. Year after year, the RSC requested that all provincial French *résidents* search for suitable recruits for

Year one students, 1924-25

Nguyen Thi Ba (Ms. Humbert)
Did not complete program

Marie Louise Canavy
Left program for family reasons

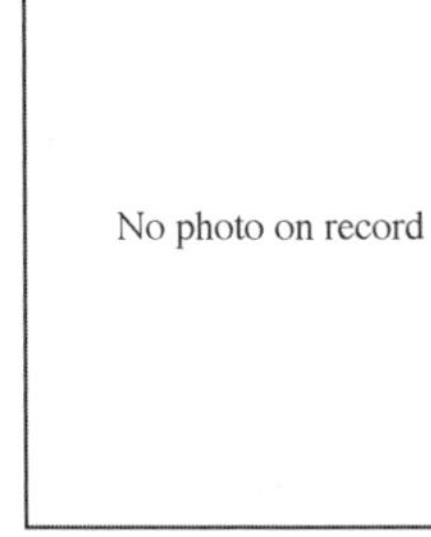


Nguyen Thi Manh
Left program to tend family business

Pham Thi Kim
Worked to 1937;
quit after bribery scandal

Saroun
Tried to quit immediately after
program, but had to work to 1932

Suos Kuong
Dismissed from program
in first year

Suon
Completed program

Tran Thi Huon
Completed program

Figure 5.9. First three years of admission to École des Sage Femmes, Phnom Penh. Courtesy of the National Archives of Cambodia.

Year two students, 1925-26

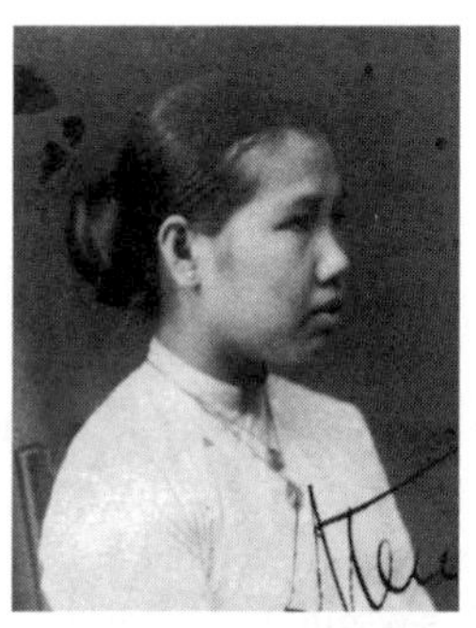

Nguyen Thi Nieng (Helene)

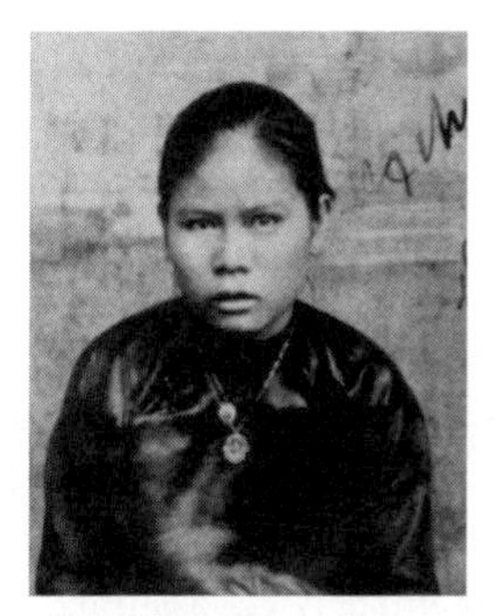

Nguyen Thi Tai

San Thieu
Resigned from program

No photo or continuing records for two other admits: Pham Toum and Phoc Lours

Year three students, 1926-27

Pauline Chea

Dang Thi Kieu
Repeated and failed second year

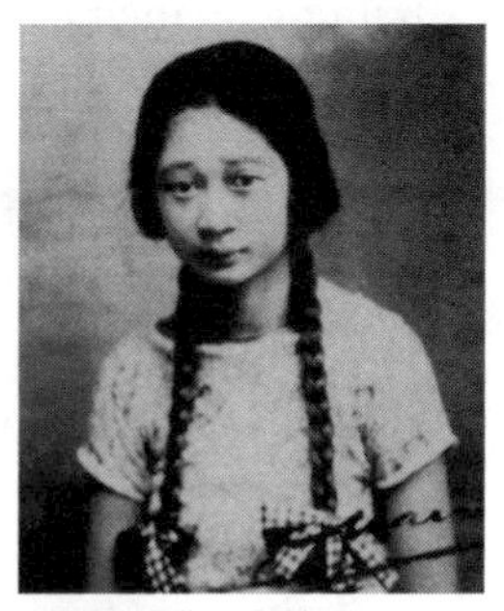

Juliette Pakiry
Left program; called home by father

Alice Reynaud (later A. Cruz)

Sanne Suos
Repeated and failed second year

midwife training. Each year, the *résidents* would reply that no suitable candidates had been found.[119] Family members expressed concern about the pernicious influences of these programs on their daughters. In one case, to convince the chief of the Chams to allow a young Cham woman to enter the program, the health director had to assure him that the maternity had no men. It was thus greatly embarrassing to the health director when, one night during training, the young woman ran off with a man. In the ensuing midnight search, the supervising staff found two coolies fraternizing with two of the female maternity staff. The health director fired all parties involved and admonished the two French female supervisors. Having kept the affair very quiet, the health services assumed that this would be the end of the matter. Unfortunately, the dismissed coolie returned to the maternity the following afternoon and attacked the supervisor with a knife. Ironically, the whole affair ended up quite publicly before the commissary.[120] Such events certainly made protective families doubt the effects of professional training on their daughters.

In 1929, five years after its opening, the school received only three applicants for the entire year.[121] Although Khmers were preferentially accepted, by 1931 only one of the five candidates found was Khmer. Khmer women who wanted to assist in birth and pregnancy seemed to prefer being trained by existing *chhmaps* in the traditional way. On December 4, 1933, the AM closed the École because no applicants could be found.[122] The one student already matriculated, Tit Saoum, opted to quit rather than enter the AM at a lower pay scale as an *infirmière*.[123] This illustrates another problem with the training of female staff: *infirmières* were also in perpetual shortage. This is not surprising, considering that they were paid considerably less than midwives, yet often performed the same duties because midwives were in short supply. However, male *infirmiers* were plentiful; men in contractual menial employment were clearly more common than women. In 1938, the Cambodian AM had openings for 22 contract *infirmiers* and two *infirmières*. It received 387 applicants for these 24 positions. This large pool of applicants included not one Khmer woman. This was also true in 1936, when six positions were available. The health director would grumble, "If the women in this country still complain about being attended to by the Annamite, they have only themselves to blame."[124]

A career in the medical service was clearly unattractive, and a survey of the careers of these women reveals why. For instance, the young Vietnamese woman Nguyen Thi Lieng fell in love with Khmer fellow student Sonn at the École de Médecine in Hanoi in 1909.[125] He was training to become an indigenous *médecin*; she was in the midwife training program. The director of the École kicked Sonn out that year for "fraternizing" with this "Annamite girl" and other "bad behavior." Deemed relatively bright, he was returned to Phnom Penh to complete probational training at the Mixed Hospital before being allowed to take his exam to become an indigenous *médecin*.[126] Sonn kept up his relationship with Nguyen Thi Lieng over the next two years. In 1911, his first posting as an AM *médecin* was in Pursat. Nguyen Thi Lieng also requested a Pursat posting from the AM, which, surprisingly, it granted. However, her aunt withdrew the request on her behalf, explaining that her niece belonged with her aged parents and young siblings in Hanoi. Her parents simply could not bear to send their "weak-spirited and fragile" daughter, who was nonetheless the "joy of the family," to the "distant, unknown country" of Cambodia. The aunt asked that the administration disregard the young woman's "extreme silliness" (*grave sottise*), her impulsive request having been made in a state of youthful ardor.[127] In internal discussions, the secretary of the École wrote (rather unfairly) to the Cambodian Health Director that he should simply fire Nguyen Thi Lieng for her inconstant character. The secretary continued, "It would be better if Sonn married a Cambodian, because he would in this way gain greater influence over his peers. I would also add that thi-Lien [sic] does not appear to have sufficient guarantees of morality."[128] Sonn's morality was not at issue. The AM ultimately granted Nguyen Thi Lieng's or more accurately her aunt's request to remain in Hanoi. Several years later, Sonn would marry the ethnically appropriate Neang Tiès and have a very successful career with the AM, becoming one of the most prominent Khmer doctors in the colonial and postcolonial period.[129] In contrast, family pressures ultimately forced Nguyen Thi Lieng to put aside her career plans and personal wishes to remain with her parents and siblings. While a man could fulfill his familial obligations by earning an income, a woman's social obligations were much more limiting. Nguyen Thi Lieng and these other "career" women were attempting to create a new public social role. The health services would continue to show little sensitivity to the conflicts that these new obligations created.

The women in the medical services were under stress similar to that experienced by the men, except that their precarious positions were exaggerated. The stress could be considerable. A 1933 graduate of the École, Dang Thi Nuoi, had a nervous breakdown within two years and was fired after her supervisor confirmed her physical and emotional instability.[130] Several other women also left for reasons of fatigue and ill health.[131] Consider also Nguyen Thi Hieu, sister of fellow midwife Nguyen Thi Chung, a young woman who died of unspecified causes shortly into her career. Unable to resign and refused a leave of absence, she abandoned her job without notice shortly before her death. At her death she was still technically absent without leave, and owed the administration 464 piastres in school fees.[132] Class and connections could make a difference, as the example of Alice Reynaud reveals. Reynaud entered active AM service in 1928 (figure 5.9). Her story contrasts with many others due to her financial solvency. She married a European, and because she was able to pay back student fees, the administration gave her leave in 1930 (by then she had taken the married name Alice Cruz) to take care of her children. She resumed active work with the AM in 1938, when her children were older.[133]

Perhaps it was simply a pretext, but family members "called home" several of these women in mid-training to fulfill family obligations.[134] Further, after graduation, the administration posted women regardless of their preferences, and several quit when they were assigned away from their husbands.[135] Unlike the men, these women were under a social expectation to follow their spouses. Thus, we may see complaints by the male medical staff about constant shuffling of their postings, but all records indicate that their families either stayed in Phnom Penh or followed the men. There is only one recorded case of a man requesting a changed assignment to move to his wife's posting (Nguyen Van Ngoc, discussed below). Because so few outlying districts actually had dedicated maternities, the possible postings for women were much more limited. Nonetheless, wives almost exclusively requested reassignment to join their husbands. Even if the woman had no family obligations, constant relocation was all the more discouraging because women did not usually travel on their own.[136] The problem was exacerbated because the women in these programs usually had family members who were employed as French civil servants. On the other hand, the number of men in professional careers greatly outnumbered women, and many men in the medical services had wives who were not employed by the French. They were perhaps housewives or merchants. Ultimately, the career and personal life of the wife took the brunt of these administrative shufflings.

One of the most tragic examples of the cost of these conflicting pressures can be found in the story of Neang Saroun, a student in the inaugural year of the Phnom Penh midwifery school, who entered the program at fifteen years of age (figure 5.9). When Saroun successfully graduated in 1927, she married a young Khmer man who was a *secretaire des résidences*. In her first year of marriage, the colonial administration posted her husband in Siem Reap and placed her in Kompong Chhnang.[137] Thus, immediately upon entering AM service, she requested a reposting to Siem Reap and, barring that, temporary leave for one year to be with her new husband. Her request was denied in part because Siem Reap had no maternity hospital.[138] The husband did not request relocation to Kompong Chhnang. In refusing her request, the health director wrote

> Before Miss Saroun requested entry into the cadres of the Assistance in Cambodia, she knew perfectly well that the needs of the service might designate her for a post not of her choosing. . . . It seems unfair on the part of this midwife to seek all the advantages that come from the position of a civil servant without accepting the burdens.[139]

Because she was unable to repay the educational costs at that time, Saroun reported to her assignment in Kompong Chhnang. In 1930 she again requested temporary unpaid leave, explaining, "I find myself in a situation that does not permit me to reconcile my job as a midwife with my duties as a mother, and strongly desiring to breastfeed my newborn myself and care for my other small child, I very respectfully request one year of leave."[140] Again the administration refused her request, informing her that she would need to immediately repay all school fees if she failed to serve her six years of contracted service. The strain on Saroun must have been considerable. Posted in Kompong Chhnang, she was away from both her husband and her wider family network (in Phnom Penh). Two years later, at the age of 23, she died of unspecified causes, leaving behind three small children, one only a few months old.[141] The social expectation to have children and the need to care for infants at a time and place with no child-care facilities conflicted with Saroun's duties to the AM, and likely contributed to her early death. Considering that these women were being recruited to educate indigenous mothers on the correct care and nurturing of their children, the inflexibility of the medical services with regard to Saroun's maternal needs was deeply ironic. In this particular case it was also tragic. Such AM bureaucratic indifference doubtless served as a deterrent to other entries into the service.

## The Métis in the Sage-Femme Programs

Another striking aspect of the sage-femmes programs was the prominent role of the métis. The health services had a difficult time finding indigenous women who met the prerequisites for the training programs. Entry requirements included extensive paperwork that required a working knowledge of French bureaucratic procedures. The educational stipulations alone excluded most Khmer women, because few of them attended school in the colonial period. Even fewer had attended a French school long enough to speak French proficiently, yet another requirement of the training programs.

There was, in fact, only one substantial pool of qualified applicants in all of Cambodia: the female students of the SPE. The SPE took the moral and intellectual guardianship of its students very seriously. The board of the SPE wanted its métis charges, particularly males, raised in France to lose local influences "pernicious to their mentalities." Due to lack of funding, this was often not possible.[142] Instead, the organization educated and molded its students locally with the resources it had available. Women particularly were to learn the correct manners and behavior necessary to be good wives. In 1915 the director of the Cambodian SPE, Gravelle, claimed that the organization had considerable success in educating its male wards in practical schools and finding the women suitable matches: "We apply ourselves in marrying our young women, due to the lack of Frenchmen (very rare), with Asians, Annamites or Cambodians able to make them happy.—We want them to be honest homemakers, always protected by us, bringing into the indigenous milieu familiarity and sympathy for the notion of France."[143]

The RSC expressed a keen interest in the SPE's work, particularly as it dealt with potential health staff. In a letter to the governor general discussing the development of the health services in Cambodia, the RSC quoted at length several of Gravelle's published statements in a colonial newspaper. One such excerpt discussed the future direction of the SPE female wards.

> Until now, within all the organizations formed for the protection of abandoned métis, education, despite the best intentions of the founders, seems to have been neglected and the young women's school has created more "demoiselles de Compagnie" than good housewives. . . . My opinion is that it is advisable to fix without delay such a state of affairs by giving the métis a carefully watched professional training and an education before we will be able to change the mentality noticeable among them.[144]

The RSC clearly recognized quite early on the utility of this group. Thus, it is not surprising that a disproportionate number of entrants in the Phnom Penh midwife school were students of the SPE. In the first year two of the eight candidates were SPE métis: in the following two years, four of the ten candidates were métis. This pattern would continue in later years.[145]

The métis women entered the AM in the indigenous cadre. However, it was to their financial advantage to convert, if possible, to the European cadre with its higher pay scale. In some instances, racial pride drove attempted conversion. In most cases, both financial and racial considerations likely motivated these attempts. Other métis, not necessarily recognized by the SPE, also became prominent within the program. Again, this was due to their culturally and racially liminal state. They were both French and Indochinese, and neither one completely.

For instance, Nguyen Thi Chung, also known as Agnes Nguyen, was a young métis woman who entered the Cambodian AM in 1918. Having been raised by her indigenous mother, she trained at the Hanoi École de Médecine.[146] Both she and her husband, Nguyen Van Ngoc, an indigenous *médecin* who entered the AM in 1921, were initially posted in Phnom Penh. In 1923 the health services transferred him to Battambang. Remarkably, although Agnes was also given the choice of reassignment to Battambang, she opted to remain in Phnom Penh and unsuccessfully attempted to have her husband reposted there as well. Since he could not return to Phnom Penh as an AM employee, he went on unpaid leave two years later, and finally resigned from the AM in 1928 to concentrate on "entrepreneurial undertakings" in Phnom Penh. Although he had waited six years to resign, the administration ruled that technically he had not worked for six years, and forced him to pay back 1,220 piastres in educational fees.[147] This debt may have played a part in Agnes's attempt to convert to the European cadre with its higher pay rate. In 1931 a court decision recognized her French citizenship—necessary because her father had died without ever acknowledging his paternity. Agnes then lobbied for several years to move into the secondary European cadre of the AM. In one of her many petitions, written in 1937, she wrote, "French myself, I have suffered and worked for 18 years with the shame of being denigrated by my white peers. I have not been able to cross the barrier that separates me from . . . those who belong nonetheless to a family to which I am proud to belong, the great French family."[148] In this year, the GGI finally allowed her to take an examination to switch cadres. Whether she passed the exam is unclear, but by this point the issue was only a matter of pride. Agnes retired two years later with a government pension.[149] The difficulty in her converting to the European cadre had resulted from her lack of SPE sponsorship. The SPE assisted its wards in all aspects of acquiring "Frenchness." Its council, as a matter of course, aided students in legally obtaining French citizenship in exchange for "their good conduct and hard work, proven [in part] by obtaining one of the diplomas delivered by a technical school." As punishment, the students who gravely misbehaved were "returned to their indigenous families."[150]

Agnes Nguyen was one of the few midwives, métis or indigenous, who worked for the AM consistently and for any length of time. Métis women may have had slightly more academic success than their Khmer peers, but they ultimately proved as unreliable to the medical service as were the "fully" indigenous women. The constant shortage of female hospital and maternity staff was not alleviated during the French colonial period. In a way, this was immaterial, because indigenous avoidance of the hospital and the maternity also did not appreciably decrease in the colonial period. The constant shortage of staff may have caused the shortage of patients, but a more plausible explanation would seem to be that both shortages were simply different manifestations of the same phenomenon. Indigenous women had little desire to interact on a contractual basis, either as patients or as employees, within the colonial apparatus. In the mid-1930s, the AM shifted its plan of action.

## Mohammed and the Mountain

In 1936, two women who had been coerced into taking an entry exam for the newly revitalized midwife-training program disappeared after passing the exam. In light of these desertions, as well as the lackluster results in recruiting both female staff and female patients, the health director outlined a new tactic in gaining medical access to women: "Because the Cambodian woman will not come to us, *it is necessary to go to her, through her.*"[151] Going "to her" involved sending out medically trained staff into the rural villages. Going "through her" meant recruiting a large number of young women from these remote villages, training them, and sending them back to their villages to perform scientifically informed births and disseminate correct child-care methods to the wider population.

The idea for this "rural birth attendant" (*rural accoucheuse*) program had been steadily evolving from other social outreach programs. The Pursat maternity, which opened in 1927, had found the number of hospital births growing at a discouragingly slow rate. In an effort to increase outreach, the health chief implemented a program of recording recent births. Midwives then visited homes with newborns, counseling new mothers and educating them on child rearing and health in pregnancy. These efforts were intended to influence these womens' entire villages, thereby reaching otherwise unreachable segments of the population. The health director also wanted the numerous older *chhmaps* in the countryside to be regulated. Although this regulation was never implemented, he had hoped that it would eventually lead to their elimination and replacement with diploma-holding midwives.[152] These suggestions were likely inspired by the visiting-nurse model being adopted in the French countryside during this decade. Again, the metropole was the experimental proving grounds for programs later attempted in the colony. This model can be traced back further to the American Red Cross, which brought the "American" model of nursing care developed in the antebellum United States into France during World War I.[153] In 1931 the health director began musing about sending midwives out systematically into the countryside, writing, "There is a place to foresee the augmentation of the number of auxiliary midwives and equip all of the important provinces with them, since Cambodian women, because they are afraid, do not present themselves to the doctor."[154] This would be the premise of the rural birth attendant program.

## The Rural Birth Attendant Program

The AM instituted the rural birth attendant program, modeled most closely on a program in neighboring Vietnam, in the districts of Prey Veng, Stung Treng, and Pursat in 1936.[155] In 1937 the program expanded to the rest of the country. Wide regional variations existed in the success of trainee recruitment for the program. Young recruits went for six months' training at the Roume maternity, after which time they were returned to their hometown. The administration paid them half a piastre for each birth attended in the countryside. Remuneration was not substantial in part because the AM anticipated that the traditional gifts provided by the birthing mother would supplement government income. Due to the untraditional structure of the program, birthing mothers did not end up providing these anticipated traditional gifts. The Prey Veng *résident* was exceptional in providing the attendants free housing and a quantity of rice. This no doubt helped explain the district's relatively high recruitment rate of fourteen trainees in 1936, compared to two from Pursat and none from Stung Treng, despite "considerable effort" on the part of both medical chiefs. The Pursat doctor believed the program was too radical. He recommended to the health director the alternate course of training preexisting traditional midwives, the *chhmaps*, rather than new ones; these retrained *chhmaps* would then gradually influence the next generation of birth attendants and mothers.[156] The Kompong Chhnang doctor observed, somewhat pessimistically, that *chhmaps* were too numerous in his district. Further, since this type of work was seen as neither lucrative nor desirable, and was usually hereditary, he believed potential trainees would be hard to find.

The Kompong Chhnang medical chief would seem to have been largely correct in his assessment of local receptivity to the program. In 1938 he would applaud the eagerness of his new rural birth attendants, while cautioning that their new roles were indeed clashing with preexisting traditions: "[A] real institution [of the program] would have certain difficulty in blending with rural customs, because, in part, it would upset traditions and, on the other hand, the rural birth attendants, almost all very young, hardly inspire at first sight the same confidence in the population that the old matrons inspire in them."[157] Although these "old matrons," the *chhmaps*, did not enjoy great material prosperity, they seemed to command considerable respect and appreciation in their communities.[158]

By 1939 the country had ninety-one rural birth attendants, an average of twelve per province. The number of births these attendants reported was impressive. For 1938, these women reported attending nearly thirteen thousand rural births.[159] However, the new health director that year was, unlike his predecessor, skeptical of the program. He wrote, "The greatest inconvenience of this institution is the absolutely uncontrollable character of its functioning. And it is not possible to consider the numbers furnished regularly from certain provinces as anything but fantasies."[160] He would also observe that the previous rural birth attendant program on which it was modeled, the program in Vietnam, was introduced to reduce the high mortality from umbilical tetanus, an almost nonexistent problem in Cambodia.[161] In sum, the rural birth attendant program had several foreign antecedents. Its structure had been initiated in the antebellum United States, it had been developed in interwar France, and its justification had come from neighboring Vietnam, and yet the health services were puzzled about its ineffectiveness in the Cambodian countryside. As history would have it, the program did not exist long enough to have a discernable impact on the rural population. As with so many of the medical programs in the country, the rural birth attendant program dissolved with the outbreak of World War II and the ensuing governmental instability.

## Pregnancy and Medical Intervention

The reasoning and motivation for French medical efforts towards women's health can be summed up briefly. French doctors wanted to improve population health; they would ensure that children were born correctly, and properly trained midwives would ensure that they were properly nurtured. Khmer women, however, were never convinced that their children had been born incorrectly or cared for improperly. Again, there was a gap in communication between doctors and would-be patients. French actions aimed at changing behaviors never addressed the motivations behind those behaviors. They could not, because they did not understand these motivations. Khmer women may not have comprehended the reasoning behind French medical efforts, but they did understand their own lack of control within the maternity hospital. On some level they also must have understood that their own concerns and very real fears were not taken seriously by their caretakers.

Along with the change in attendants, the change in locale demanded by the French was significant. As one study in the European context observes, "the place of birth shapes the experience, determining who is in control, and the technologies to be employed. In a home birth, those attending are visitors in the family's domain and . . . must rely on the family for an understanding of local customs and practices."[162] At the maternity hospital, in contrast, the doctor has the power and determines what will be practiced. However, these negative characterizations must be tempered by a realization that social control was not the only motivating factor in colonial involvement. Further, traditional birth rites with an attendant *chhmap* also involved a degree of social control and conformity. Just as important, the government's maternalist attempts also facilitated the first entry of many Khmer women into public and political engagement with the colonial state. In the same vein, French offers of a different sort of medical care, while paternalistic and authoritarian, also provided health alternatives that could be greatly beneficial to the patient.

The transformation of pregnancy and early motherhood into a medical matter has been rapid and profound for most Western nations in the twentieth century. For Cambodia this transformation did not occur during the colonial period, and it can be argued that it still has not occurred, particularly in the countryside. Childbirth is still predominantly a ritual event centered in the home.[163] This lack of transformation may be a failure in the expansion of Western medicine, but is it a failure in terms of improved health for women? In truth, we cannot accurately identify "women's health" itself as the ultimate goal of any of the French medical efforts towards women. In the West, society *as a whole* was undergoing various social changes—the increased authority of science and medicine, the shift to population thinking, the insertion of technology into daily life, and so on—that encouraged the radical transformation of women's reproductive health. This was not so in the colonial state. Although the French colonial government clearly shifted its thinking on reproduction, once again the indigenous population failed to follow suit, unlike in the industrialized West. This is say not that indigenous understanding of childbirth was static, but merely that the social and cultural changes in Europe facilitated the shift in the role of medicine in women's health. The social changes in the colonial setting, also radical, followed rather than facilitated an attempted change in technological practice.

The relationship between female empowerment and the maternalism movement in the colony is a more confused story than that of medical practices in Cambodia. Some feminist authors argue that pregnancy and birth are natural and normal rather than inherently pathological, and that these events should not be under the purview of medicine.[164] As for postnatal or antenatal care, feminist writer Ann Oakley argues that medical involvement with this antenatal care "is both an exemplar and facilitator of the wider *social* control of women."[165] The problem with such analyses is that they can overlook the fact that as the womb becomes politically significant, so does its bearer. In Cambodia, the perceived need for access to the womb created an opportunity for women as care providers, as patients, and as participants in a public debate. It is true, however, that most women in Cambodia did not perceive these new roles as new opportunities.

The failure to recruit women as medical care providers can be considered part of the same phenomenon as the failure to recruit women as patients. The limited role of Khmer women in the public sphere was key. Women did not gain much more political or economic power through these new roles created by the health services. A useful model for analyzing gender relations in Southeast Asia is that proposed by Norma Sullivan in her study of gender-roles in contemporary Yogyakarta, Indonesia.[166] Sullivan found that "as a generalization, it is fair to say that the private world is female-directed and the public world male-directed, and that the former is reasonably styled a 'female world' and the latter a 'male world.' It is also clear that the former is ultimately subordinate to the latter."[167] The woman "manages" a specific domain over which the man is ultimately the "master." In Sullivan's view, modern nationalist discourse and scholarly historicization of an ideology of "a traditional separate but equal" woman's role in the domestic sphere serves to further circumscribe women's power in the modern era. The social constraints clearly at work in limiting the medical career of Cambodian women demand that we approach the concept of precolonial gender equality with care. Also, the degrees of sexual egalitarianism also varied greatly among different Southeast Asian ethnic groups. Khmer women, by many accounts, had a much more constrained public role than Vietnamese women, for example.[168] Although women may have enjoyed "high female autonomy"[169] before the colonial period, colonial modernization—even as in some senses it commodified women's reproduction and productive capacities—enabled a new public space for women.

As Barbara Andaya has observed, historians have done little in "comparisons of the way in which language and images associated with child-birth and child-nurture were parlayed into the actuality of human relationships across different cultures and different belief systems."[170] We still have an incomplete understanding of how Khmer women perceived their maternal roles and how their perceptions changed during this period. However, from some of the vignettes in the historical record, we can get a sense of mothers' conflicting obligations under the colonial medical administration. Because the "traditional" role of women in Cambodia before the colonial era was so tightly constrained in the domestic sphere, the careers of the women we have examined cannot be considered separately from their personal lives. The public roles that were being foisted upon (or offered to) these women by the colonial government were often incompatible with the existing roles they were still expected to fulfill. Their positions were made difficult not only by the insolubility between different cultures of childbearing, childbirthing, and childrearing, but also by conflicts in cultural perceptions of gender relations and professional obligations.

In Europe and America, a growing state focus on reproduction became quickly entwined with a maternalist movement in which many "middle-class women viewed motherliness not as their special burden or curse but as a peculiar gift that encouraged them and justified their efforts to gain some measure of personal and political autonomy."[171] The control of motherhood empowered these women to lobby in the public sphere for women and children's rights that extended beyond the male-controlled state interest in population and reproduction. However, women in Cambodia had little opportunity to shape or temper women-focused policies according to their needs. This is also in contrast to the situation in France. For example, women in France successfully organized to refocus the interest of the state from simply increasing the birth rate to improving women's general health. As a result, in 1913 French women were required to take maternity leave for the health of their children (and, by extension, the nation), with the state partially subsidizing lost wages.[172] In the colony, in contrast, the maternalist movement largely imitated European trends, with European women creating maternalist organizations to assist their "indigenous sisters" who would play only a marginal role in their direction.

While in Europe women actively politicked for maternal and children's political rights, the maternal movement in Cambodia was only an initial step in creating a political engagement between indigenous women and the state. This step may have been small, and the positions created may have conflicted with preexisting social roles, but they also created new opportunities.

In other colonial contexts, scholars have revealed how indigenous women have used medical careers as a door to political engagement.[173] For the women discussed in this chapter, it would be too bold to claim that they obtained a political voice within the two decades of this study. The programs were ultimately too new and transitory. However, those women who did participate, even fleetingly, within the medical system engaged with their government employers. They were actively negotiating a new role for themselves in a new domain—moving from a domestic into a public sphere.

SIX

# Civilized Lepers

A theme throughout this study has been the differing epistemologies in French and Khmer cultures surrounding medicine and disease. These epistemologies underlie the social definition of a disease, which in part determines its appropriate treatment. They also concern more broadly the proper understanding of disease, politics, social behaviors, and moral order. The debates surrounding leprosy also provide a window into the complex negotiation between the particular nature of human experience and the universal aspirations of science.

Through the story of leprosy, this chapter makes two arguments. First, it contends that the colonial government's vision of leprosy in Cambodia involved a nascent epistemological conceptualization that combined preexisting aspects of leprosy's social taint with the new science of laboratory medicine—and that this vision was driven by France's history separate from Cambodia's specific environment. Like degeneration and depopulation, leprosy was a social problem defined in the metropole and transferred to the colony.[1] Two redefinitions were occurring. In the colonial context, the Cambodian social construct of a leper was unsuccessfully overlaid by a French construct. But within biomedicine the disease was being reenvisioned through the concept of a bacteriological agent, and this new vision was accompanied by a new disease nomenclature. This renaming served to remove the wider social stigmas of certain afflictions while standardizing the medical nomenclature of specific bacterial conditions. Thus, doctors at the turn of the twentieth century increasingly referred to leprosy as Hansen's disease, in honor of the first man to identify the leprosy bacterium. Physicians assumed that the connotations of Hansen's disease could be circumscribed to the bacterial infection; the label leprosy was supposedly much less objective and more stigmatizing. Yet, these "less objective" connotations were in part Western imports. Ironically, colonial scientists introduced the "leper as unclean" into Cambodia at the same historical moment as medicine more broadly was trying to efface just this stigma.[2] These two processes, neither linear nor mutually exclusive, were often in direct opposition.

Further, the leper's transformation into the Hansen's disease patient was strongly affected by the relationship between the laboratory and the biological entity. Leprosy is one of modern medicine's most intractable mysteries, for the *leprae* bacillus, having been discovered relatively early, resisted culturing in the lab.[3] Scientists could not grow it in vitro, and thus could not perform on it what were becoming standard microbiological experiments at the turn of the twentieth century. Its method of transmission was, and still is, unclear.[4] Although coupled to a bacterium, leprosy's biological impenetrability made it an exceptional affliction that noncontagionists could invoke to support older hereditarian views of disease etiology. One of Western society's first Others also became one of medical science's first Others. In this chapter we will examine how different social groups negotiated among these social, scientific/laboratory, and biological attributes of leprosy.

Also, the story of leprosy management starkly highlights one of most significant factors that limited the success of French medical care in Cambodia. As mentioned before, colonial administrators attempted to change leprosy's social definition in part to alter the perceived suitable treatment. This change was enacted from the top down, from the level of public health specialist to the general administrator to the colonized subject. Yet indigenous society, including both the afflicted and his peers, conserved a different meaning for leprosy, as is captured in the Khmer term *chomngu khlong*. More clearly than for many other diseases, the battle over this definition can be read in the figurative uses of leprosy in the spoken and written language.[5] The discursive battle is reflected in the institutional history of leprosy, which further illuminates the complicated struggle over its social metaphors. Just as the utter failure of leprosy's rationalized management highlights the paradoxes in the universalized ideals of efficient scientific and governmental control, contradictions in its changing local and European definitions reveal the contingent nature of a single, universal, scientifically constructed disease identity.

## Transgressions

Within Western literary and religious traditions of the past two millennia, leprosy has been associated with profound ethical and moral decay. The word *leper* evoked both a disease and a state of social exclusion, both patient and sinner.[6] When officials in the colonial government spoke of leprosy, they were working from this construct and its European associations. In one of the earliest notes on leprosy in Cambodia, dated 1898, the medical chief of the Mixed Hospital suggested the creation of a leprosarium[7] in the vicinity of the Catholic Church in Phnom Penh. All lepers who could not be "returned to China or the provinces" were to be placed with the Sisters of Providence, who were "very willing to be involved."[8] This locale for interning lepers was only natural if considered in the Christian framework that linked leprosy, sinners, martyrs, and the Church.[9] While Christian charity did extend more broadly than the Christian notion of sin, the Church found leprosy to be one of the most effective diseases evoked to solicit colonial missionary donations.[10] It was no coincidence that a century earlier, in 1802, the Vietnamese emperor Gia Long had organized the first leprosaria in the territory with the assistance of French Catholic missionaries.[11] The first Indochina leprosarium, Culao Rong, would also have a strong Catholic missionary presence.[12] Care and isolation of lepers was initially conceived as a Christian task, but it also would become a scientific task.

While the link in much colonial writing between the leper, moral transgression, and religious redemption was inescapable, such associations were rarer in the Khmer context. The small volume of any secular writing in Khmer at the turn of the century limits the indigenous literary representations of leprosy in the historical record. Traditional tales of the leper in Cambodia provide alternative visions for mediating experience with the disease. From these stories we can glean various cultural readings of the leper. The most popular of these tales was the story of the leper king, which in short details a great Khmer king who becomes afflicted with leprosy after a mandarin subordinate spits in his eye. A monk attempts to heal the king, but the king fails to show him respect. From this act, he becomes permanently afflicted.[13]

This story is likely a bastardized folk memory of a true historical figure, a real "leper king," and thus any figurative meanings extrapolated must consider its historical basis.[14] However, the centuries-long retention of the tale in popular culture, along with the commonality of specific mythic elements, intimate certain Khmer cultural readings of the disease. This tale indicates that leprosy is seen as caused in part by moral transgression, but also by physical taint. Although the original transgressor—the mandarin—could be seen as a subordinate, the disease itself afflicted a revered figure. The etiology of leprosy, like that of many other diseases, is supernatural. Thus, much as in Western constructions of the disease, transgression leads to affliction. However, in this story the transgression is social as much as moral; both disrespect and unworth determined affliction. The disease did exceptionalize, but it did not necessarily denigrate (reduce the worth of) the sufferer, as it did in Western Christendom.[15] If anything, Khmers seemed to view the leper with cautious respect. The leper king was seen as cursed but still powerful and dangerous.[16] The tension among the three levels of power in this story (magical, meritorious, and royal) is a common motif of Khmer folktales.[17] The story is also reflective of the belief that moral order and world order are intricately linked, and that right action (in the ethical as opposed to pragmatic sense) determines social status. The story's popularity suggests that it provides a normative view of specific disease-related behaviors.[18]

A semi-inclusive status for the leper in indigenous society is also supported by foreign observation. In the thirteenth century, Tcheou Ta-Kouan, a visiting Chinese mandarin in Cambodia, commented, "Even when they [lepers] come to sleep with us, to eat with us, the natives do not object."[19] A colonial physician echoed this sentiment seven centuries later, saying of the effect of lepers on the Indochinese, "They inspire neither fear nor repugnance from the population."[20] Time and again, European commentators, sometimes greatly puzzled, observed that Khmers simply were not concerned with catching the disease, intermingling rather freely with their disfigured brethren.[21] Leprous merchants openly traded at the markets.[22] The story of the leper king, as well as these observations, indicate that the leper in Khmer society was on some level stigmatized, but unlike in French society, segregation usually occurred within the bounds of community.[23] Significantly, today a statue of the leper king at Angkor Thom is still treated with great reverence by Khmer visitors.[24]

Conflicting views of the leper were most apparent when French and Khmer representations of the leper king overlapped. One such example can be found in a 1927 colonial fiction novel entitled *Le roi lépreux*, which describes a Frenchman's encounter with this Angkorean statue. The protagonist of the novel has a strong emotional reaction on encountering the stone figure of the leper king. He describes the statue: "It was of a beautiful violet sandstone, and represented a completely nude young man, of a height slightly below average, sitting in the oriental style. . . . The face, remarkably pure, had a sad nobility, almost despairing."[25] The protagonist becomes mesmerized by the image's "marvelous expression of hopelessness, of distress."[26] Although unexceptional as a work of fiction, the novel is illuminating of a particular mindset, for the facial expression of the leper king does not overtly convey "hopelessness," "distress," or "sad nobility." The despair read in his expression seems in large part creative license (figure 6.1).[27] Essentially, a French construct of a leper is read through and

Figure 6.1. Images of the Leper King at Angkor Thom. Photographs by the author.

despite the existing Khmer image. Contemporaneous with the novel's publication, the colonial administration was similarly reconfiguring the living lepers in the country. In other words, they too were being redefined over existing representations within Khmer culture.

The exceptional nature of the leper in the French colonial context is nicely illustrated by the example of a leprous prisoner discovered in Cambodia. When this inmate's affliction was diagnosed in 1911, prison authorities insisted he could not remain at the prison. After much discussion the RSC, the police, and the medical doctor transferred him to the leprosarium in neighboring Cochinchina.[28] This same leper caused more grief to authorities when he was condemned to death in 1914. The Cochinchina leprosarium did not have facilities secure enough to prevent his escape, which was more likely considering his new sentence, yet his confinement in a prison with others was still unthinkable.[29] The special consideration over this prisoner's disease and the potential infection of other prisoners is indeed strange, considering the lamentable health conditions and high death rates within protectorate prisons at the time.[30] During the same period when this single leper was carefully confined from others, active tuberculoid prisoners continued to live and die in overcrowded cells with their peers. Tuberculosis, a disease bacteriologically similar to leprosy and believed by some at the time to be more contagious, was an acceptable risk of prison life. Leprosy was not.[31] The need to confine this single leper from other *marginal* nonleprous individuals reveals that the fear of the disease was not solely about protecting European health or limiting a disease reservoir to the natives. Clearly, in the French conception, the leper was to be excluded from all levels of society for reasons that went beyond health.

## Medicalizing the Disease

An increasingly alarmist tone about leprosy, common in turn-of-the-century international medical literature, was not matched by any documented global rise in the number of lepers.[32] As with women's health (see chapter 5), leprosy was not mentioned much by colonial observers in Cambodia until it was declared a problem. The first and rather incomplete survey of lepers in the country occurred in 1897, when the naval doctor Henri-Albert Angier conducted an approximate census in the provinces. The impulse for this tour came from the GGI, which had requested leper tallies for each state of Indochina. Before this request, leprosy had elicited little more than side commentary in the official documentation. Angier, from his travels, made what he admitted to be a very rough estimate of 129 cases in the country. At the time, the population of the country was approximately 1.5 million.[33] The doctor's analysis assumed that *chomngu khlong* was leprosy, the disease that was essentially becoming delimited by the presence of the *leprae* bacterium rather than by a specific symptomology. Three years after Angier's survey, the French minister of colonies commissioned an extensive study of leprosy in Indochina. With little actual data, the author of the final report would make the alarming declaration: "The white race is not safe from leprosy. . . . If energetic measures are not taken to check the growth of this flood, there is no doubt that sooner or later leprosy will ravage the white population in this colony . . . ."[34] In 1904 the same doctor again warned his metropolitan readers of the urgent need to combat leprosy in France's overseas possessions.[35] He urged policy makers to act swiftly to "preserve [the nation] from the flood which becomes increasingly menacing with its daily expansion" throughout French territories.[36]

The number of registered leprosy deaths in Cambodia rarely exceeded single digits annually, while countrywide deaths from smallpox and cholera often surpassed a thousand in any one year. Yet the volume of colonial discussion over leprosy surpassed that for either of the other two maladies. Considering the comparatively low leprosy morbidity in Cambodia, and the even lower mortality, the concern did not arise from a pressing population health dilemma. Yet in 1908 RSC Luce circulated a questionnaire to all French administrators and medical personnel in Cambodia which began: "My attention has been drawn . . . to the notable augmentation in Cambodia of lepers of all nationalities as well as the dangers of contagion that the circulation of this category of sick individuals poses to the healthy population."[37] This "notable augmentation" was clearly apocryphal. With the exception of Angier's incomplete estimates of 1897, the colonial administration had yet to perform a leprosy census in Cambodia. Luce requested that all officials conscientiously complete the leprosy questionnaire, submit estimated numbers, and provide him with opinions on the sorts of containment and treatment to be enacted in the "interests of the population of the entire country."[38] Of the eleven opinions on record, only five were in favor of a leprosarium.

The Kompong Thom *résident*, one of the few officials surveyed who considered the opinion of the indigenous community, wrote against regional containment, arguing that none of the indigenous governors believed leprosy was contagious, and that thus they had little concern about it. Further, lepers often maintained themselves voluntarily at a certain distance from the larger villages. The *résident* opined that the RSC should follow this more economical and humane lead rather than implementing radical new programs. Self-imposed partial segregation was also much more acceptable to inhabitants than forcing the afflicted to move away, an act for which Cambodians had "extreme repugnance." In contrast to this administrator's mild view, some others, including the Phnom Penh mayor, wanted all lepers detained, confined, and "heavily guarded."[39] Certainly, the fear that leprosy inspired in some Europeans was disproportionate to its actual destructive potential. As with umbilical tetanus mortality,[40] the "Cambodian leprosy problem" was also in part a shadow cast from the greater occurrence of the disease in more densely populated and urban Vietnam.[41] Overall, the opinion of the European community was split. Those believing leprosy was insignificant pointed to its consistently low prevalence in society.[42] Those trumpeting its danger usually posed this threat in terms of its potential or imminent increase, rather than any numerical reality.

Although the opinion among the administrative and medical community may have been rather evenly divided, leprosy in Cambodia was on the route to regulation. Despite the lack of viable treatments, the scientific ignorance about its mode of action, and indeed a dearth of knowledge concerning whether it was actually contagious,[43] its rational management was deemed necessary. The mycobacterium, however, was so resistant to laboratory manipulation that a trial vaccine—the experimental panacea for all epidemics during this era—could not be developed. Further, no consistently effective curative for the affliction was known. Inspector General Edouard-Edmond Primet captured this sentiment in his 1910 report on leprosy in Indochina: "It appears to us that the practical solution [to leprosy] must be sought more from rational observations of facts and the fecund lessons of the past than from the discoveries of science, still powerless to reveal to us all of the mysteries of the Hansen bacillus, its evolution, its path of access, and its modes of transmission."[44] The identity of the disease would necessarily remain circumspect; its diagnoses and treatment were to be historically based—but on whose history?

The French medical corps did not have an effective leprosy treatment. Nonetheless, as a contagious disease, even if that contagion was as much moral as physical, it demanded isolation. To that end, on September 16, 1903, Indochina's first French colonial leprosarium opened in Culao Rong, Cochinchina.[45] To further the goal of containment, the GGI issued a leprosy decree for all of Indochina in 1909. This decree blocked foreign lepers from entering the territory and forbade existing lepers from freely circulating. This document also quite broadly banned the afflicted from any industry in which they or any objects they handled would "be placed in contact with the public." Such industries included tailoring, commerce, and food production. Lepers were suddenly barred from most livelihoods.

The GGI's decree also anticipated the creation of Cambodian leprosaria.[46] The country would not create a leprosarium until 1915, although discussions would continue over the next several years. Eventually, a preexisting Khmer leper village would be chosen by the administration as the site of governmental leper care. (We will examine the transformation of this leper village under scientific and governmental management later in this chapter.) Having chosen the route of exclusion, the government still needed to determine who would be excluded. This was not necessarily a straightforward affair.

## Seed and Soil

In the late nineteenth century, the growing field of bacteriology was integral to the nascent discipline of epidemiology. Leprosy was one of the first diseases to be coupled to a bacteriological agent, with Armauer Hansen's discovery of the *Mycobacterium leprae* bacillus in 1873. It was also considered an epidemic disease. However, it was an exceptional disease within the assemblage of epidemic diseases because of its slow development and chronic nature. Nonetheless, the one-to-one pairing, so crucial to Robert

Koch's postulates, made it a legitimate epidemic disease despite its failure to behave in a "typical" epidemic manner. The dynamics of an epidemic enabled a mode of action framed by the notion of expedience.[47] In 1902, the Ministry of Colonies included leprosy as one of the eighteen epidemic diseases requiring open declaration to colonial authorities.[48]

Just as with other diseases on the epidemic list, the Khmer population and some French doctors did not cooperate with leprosy declaration. With the occasional exception of the voluntary consultation at the hospital (often for a separate medical problem), declared lepers were individuals who warranted exclusion for other social reasons. In other words, lepers brought to the attention of the government were not declared solely because they were lepers. Rather, the disease made their social expulsion immediately possible through the colonial apparatus. Groups or community leaders often declared mendicants who were in some way becoming a nuisance. For example, the Phnom Penh Hygiene Committee combined a campaign to eliminate "unsightly beggars" (only some of whom were lepers) wandering the streets of the city with a health campaign to study the "leper problem" in 1905. The municipality elected not to immediately confine these beggars, for fear of upsetting the local population. The issue was put under further study for "both humanitarian and health" reasons.[49]

Declaration and diagnosis were two separate processes, and neither alone determined leprosarium internment. Once a leper was brought to the attention of the colonial government, medical personnel then had to confirm his diagnosis. Although the official diagnostic method was a laboratory exam of a nasal swab to determine the presence of the bacterium, in actuality a positive leprosy diagnosis was often an arbitrary process. When Indochina's first French colonial leprosarium opened in Culao Rong in 1903, the contradictory medical and cultural definitions of the disease immediately affected diagnosis. These contradictions were exacerbated by the scientist's distinct role in the colony.

The colonial researcher of leprosy was operating at a time when germ theory was just becoming widely accepted in Europe, the etiology for major diseases such tuberculosis were moving away from "hereditary essentialism" to "contagionism," and illnesses were becoming manifestations not of moral failings but of natural processes.[50] Even for the specialist, the shift to germ theory was not instantaneous. Further, the scientist needed to establish the relevance of his colonial observations to the metropolitan members of his audience; he needed to demonstrate his competence with current scientific developments at a time when contagion was à la mode. However, while conceptions of a disease such as tuberculosis slowly mutated *within* the wider French society, in the colony the doctor existed in a social framework distinct from that of his patient. Further, the colonial tendency to naturalize social and political differences within biological categories amplified racial and hereditarian views of disease causation. Bacteria may have implied contagion, yet "experience" within the colonial network encouraged an emphasis on environment, race, and heredity. Contagion itself would become entangled with French anxieties over cultural identity, race, and the disease. For example, one researcher claimed that a converted Vietnamese missionary caught leprosy simply from tending to the burial of his deceased, unconverted leprous father's corpse.[51]

These tensions are apparent in the original report on the Culao Rong Leprosarium. While classifying dirtiness and poverty as secondary causes, the attending doctor observed that the primary cause of the leprosy was always the action of the Hansen's bacteria. He would also argue, paradoxically, that a negative bacteriological test did not conclusively eliminate a leprosy diagnosis. "Clinical characteristics alone permit positive diagnosis of the disease" because the bacterium was able to "hide beneath the skin."[52] In the summary report for Cambodia issued five years earlier, the French doctor Angier had noticed that the indigenous population used a cluster of often misleading symptoms to determine the disease, and thus that many leprosy diagnoses were inaccurate.[53] At issue is not simply the direct one-to-one translation between *chomgnu khlong* and leprosy (an issue we have also seen for the plague/*rook puh* and malaria/*kron chanh*), but also the transformation of the biomedical definition of leprosy from a cluster of symptoms to the presence of a specific bacterium.[54] If the Aristotelian essence of the disease—its bacteriological identifier—could not be located, the disease's secondary presence—its clinical symptomology—still revealed its identity.

The unusual fluctuations in the activity of the bacillus, which can become dormant for long periods (a natural history which was just becoming elucidated in the interwar period), complicated the diagnostic role of the laboratory. Into the early 1930s an AM doctor in Cambodia would disapprovingly record that many lepers were interned simply on the unconfirmed[55] opinion of the doctor or an *infirmier*, despite a 1906 decree requiring two doctors' opinions and a positive bacteriological examination.[56] Such visual diagnoses continued throughout the colonial period. Yet as late as 1939, of the 578 bacteriological nasal swab exams of visually suspect patients performed at the Mixed Hospital, only 128 were positive for *M. leprae*.[57]

Once a diagnosis was made, however it was made, isolation was the primary means of "treatment." However, positive diagnosis did not always mean leprosarium internment. Class, race, and sex all played a part in determining appropriate treatment. Those with the means to guarantee their existence at home were allowed to remain under medical surveillance in their domicile. Those without were sent to a leprosarium. The guarantee of support was based on the affidavit of a prosperous relative or benefactor. The poor could make no such guarantees, and as such were forced into the leper village (although remaining there was another matter). Thus, initially, only the poor and indigent were sent to the leprosarium.

Along with class, race was an issue in internment. "Annamite" lepers were often conveyed to Culao Rong in Cochinchina rather than the Cambodian leper village of Troeung. This transfer was not only to return these men to "their country of origin" (even if they were born in Cambodia), but also because the boundaries of Culao Rong were considered less porous than those of Troeung. Annamites, "not as docile as the Khmers," were deemed more inclined to escape and thus needed stronger deterrents.[58] Yet Troeung Leprosarium was no more successful at retaining its Cambodian lepers than the leprosaria of Vietnam were able to retain their Annamite lepers.[59] In other words, Khmers were also frequently attempting escape. Further, the Culao Rong Leprosarium was as insecure as the Troeung Leprosarium. The attending doctor at Culao Rong, echoing the doctors at Troeung, would say of his charges, "One counts on their free will and their devotion to the public interest" to keep them at the leprosarium.[60] Ultimately there was not an appreciable security difference between either racial proclivities to escape or institutional locales. Lepers left when they wanted to leave and, regardless of race, many wanted to leave. The policy of "returning" Annamite lepers to Cochinchina was actually part of the colonial bureaucratic tendency to categorize subject populations by perceived racial characteristics.

Métis lepers posed a special problem. Take, for example, the métis Jean Albert Le Franc. A former student of the Société de Protection de l'Enfance,[61] Le Franc had been isolated in the Catholic village[62] for more than twelve years, his living expenses paid by the SPE.[63] In 1934, the incoming health director decided that the young man needed to be interned at a leprosarium. Le Franc's medical certificate of internment had four justifications: distinct marks on the surface of the face and ears, a large erythemato-squamous coat on the thorax and abdomen, edema and atrophy in the extremities, and a positive nasal swab for the Hansen's bacillus.[64] Three of the four points were purely visual. Having been monitored for twelve years, the young man's bacteriological status no doubt fluctuated between active and inactive. It would seem that the determining factor in his new internment order was the increased visibility of his identity as a leper, over and above his identity as a European. The RSC and the SPE exchanged several letters discussing Le

Franc's treatment options, and the SPE readily agreed to pay for his internment. Rather than Troeung, both groups opted to send him to Culao Rong, Cochinchina. However, the governor of Cochinchina ultimately refused his entry, explaining that "the leprosarium of Culao Rong, not having a special quarter for the isolation of European lepers, cannot receive the afflicted Jean Albert Le Franc."[65]

Le Franc could not be treated because he could not be categorized. The medical services were trapped within the racial categories of the colonial government. Children of mixed heritage were classified as either European or Indochinese; the métis did not legally exist.[66] The SPE, which worked to maintain white prestige, may have legally established Le Franc as a European, but this Europeanness was limited. The SPE did not have the resources to send the young leper to France for permanent convalescence. He had never been to the metropole; he could not be "repatriated."[67] The mix of Indochinese races was already undesirable within the leprosaria; a European within such an institution threatened these categories further. Ultimately, Le Franc was neither European enough nor Indochinese enough for the treatment appropriate to either "race." The archive does not indicate his fate.[68]

Indigenous women were also rare at the leprosarium. In 1920, only ten of the fifty-four lepers at Troeung were women.[69] The imbalance in this sex ratio would remain as the total number of lepers increased in the interwar period.[70] This was also true in neighboring Vietnam.[71] The reasons for the low ratio of women to men was in part the same for the leprosarium as for other areas of French medicine. As mentioned in chapter 5, women rarely came under French medical control. Doctors realized that females were underrepresented but, as with medicine more generally, they were at a loss as to how to gain medical access to the female body.[72] However, leprous indigenous women served as a focal point of various colonial anxieties, particularly since some scientists believed that leprosy, like syphilis, was transmissible as a venereal disease. In 1890, a researcher in Tonkin provided three "case studies" of Europeans who had developed leprosy through contact with the indigenous population. The main cause of contagion, he argued, was that each of the afflicted had cohabitated with or married an indigenous woman. He reached this conclusion despite two significant complications that he briefly noted: the "apparent" continued health of one of the "creole" wives after the husband developed the disease and the previous employment of one of the afflicted Europeans as a doctor at a leprosarium.[73]

Consider further a 1903 research article about the lepers of Indochina. Clinical exams on Tonkin leprosarium inmates led this researcher to conclude that negative visual diagnoses of native women were not sufficient guarantees of health. Even should the leprosy bacillus be present in the vaginal mucus, these Annamite women frequently showed "no signs of inflammation on their genital organs."[74] The author warned, "there are those who infect the female genital organs with the Hansen bacillus, which then can linger and possibly develop saprophytically without showing . . . any local manifestations."[75] The lack of visible signage enforced the need for a vigilance necessary to maintain bodily integrity. This sense of defending the self from the other also included separating healthy from diseased, and European men from indigenous women. And again, this scientist was negotiating between the clinical/visual and the laboratory/bacteriological claims to knowing the disease.

Naming the disease and isolating the afflicted were both contentious affairs in the colonial context. Creating boundaries—whether between whites and Asians, sinners and innocents, or the diseased and the healthy—was certainly a large part of the story of leprosy, but not the only part. Although leprosy was much more than a disease, a disease rested at its core, and modern medicine's goal was not simply to prevent but also to cure. Thus, experiments on potential cures also played a vital part of leprosy management. The story of these attempted treatments are also rich with figurative representations of the disease filtered through the colonial condition.

## Treatment

The most common treatment for leprosy in the French pharmaceutical repertoire was chaulmoogra oil.[76] Copious injections of chaulmoogra were the generally accepted, if not terribly effective, treatment for the disease in many European colonies. In the early period of French colonial imposition, researchers showed strong interest in indigenous Khmer remedies of possible efficacy. Scientists in the late nineteenth century evinced open curiosity about a variety of indigenous medicines, even as they presented scientific medicine as superior. With the development of the AM, the appearance of Western medicine's superiority became necessary to buttress colonial medical intervention, and open scientific interest in indigenous remedies became rare. At the same moment, the revolution in germ theory spurred an acceleration of international biomedical research, and as a result medical technologies rapidly improved in the late nineteenth century.[77] In a sense, the reality of Western medicine's efficacy moved closer to the illusion. For leprosy, however, the real efficacy of scientific medicine did not catch up to the illusion of its superiority during the colonial era. Further, the chronic nature of leprosy shattered the temporary illusions that did exist, perhaps in a way that the unsuccessful vaccines for cholera and the plague—due to the transient nature of these diseases—could not. Unlike the choleric, who either died or improved, the leper was a living reminder of the ineffectiveness of treatments offered by the colonial government. Thus, for leprosy more than for any other disease, and for longer into the colonial era, French administrators were open to experimenting with indigenous solutions.[78]

Possible treatments from neighboring Vietnam as well as further afield were seriously considered by the Cambodian AM.[79] Not only the Troeung doctors but also other AM medical chiefs collaborated with Khmer healers in their treatment of lepers.[80] In Cambodia a variety of indigenous treatments existed for leprosy. Traditional healers used an assortment of herbs, charcoal, and minerals, as well as magical rituals.[81] Magical wards served as both preventative and treatment; most Troeung lepers had distinctive tattoos, many of which were likely intended as protective.[82] The most promising of indigenous treatments was krabao, the Troeung medicine of choice.[83] Developed by Troeung's founder, Kru Pen, who will be examined in more detail shortly, this krabao treatment seemed to hold tremendous curative promise. By 1923, after several Troeung patients tested biologically negative for leprosy, Pen's successor Kru Pok was called to the royal court to present his treatment. The health director was impressed enough by the evidence to promote further research in 1923, and the AM began extensive experiments on krabao.[84] Injections of chaulmoogra were contrasted with krabao and other alternative remedies through the 1920s and 30s.[85] Although many results were published, the findings of these various experiments were never conclusive.[86] Massive ingestions of krabao according to Kru Pen's preparation methods would remain the predominant method of treatment at Troeung throughout the colonial period.[87] In the interwar period, new chemical treatments were tried. In 1924, patients in Kampot were subjected to high-dose injections of cacodylate soda.[88] By the 1930s, the Mixed Hospital staff treated lepers with methylene blue injections, although doctors readily employed krabao if the patient specifically requested it.[89]

Even with the development of new chemical treatments, the scientific and commercial interest in krabao remained strong through the 1930s. Being a remedy specific to Cambodia, it held the potential for being a "French discovery" for a leprosy cure to be used throughout the world. The prestige and economic importance promised by such a discovery garnered some diehard supporters of krabao experimentation among the colonists in Cambodia. Despite this interest, no truly convincing leprosy treatment was discovered during the colonial era. This did not stop the medical services from rounding up lepers, ostensibly to offer them treatment. As a history of the leper village of Kel Chey (which later became Troeung Leprosarium) reveals, medical "treatment" was not the primary factor in leprosy management in Cambodia.

## A World Somewhat Apart

In late 1907, during a routine smallpox vaccination tour in the province of Kompong Cham, the AM medical doctor Menaut encountered a severe thunderstorm. Traveling on horseback along an oxcart trail about fifteen kilometers from the provincial capital (also named Kompong Cham), he fled along a small pathway running away from the trail into the forest.[90] Some four hundred meters along the path, he took shelter in a large, empty straw hut in the center of a small village. At the storm's end, the inhabitants came out to investigate the stranger. At that point Menaut realized to his astonishment that every single villager was a leper, and the hut in which he had taken shelter belonged to Kru Pen, the founder of the village. When Menaut asked to meet Pen, a villager volunteered that he was probably in the forest seeking medicinal herbs. Menaut later discovered that this villager was Pen himself.[91] Pen, it would seem, was initially distrustful of the outsider.

At the time Menaut stumbled upon the village of Kel Chey, it had twenty-six lepers originating from throughout Cambodia, drawn by word of mouth to be treated by Pen, who was rumored to have a cure for leprosy.[92] Pen is one of the few indigenous healers of the colonial period whose personal background is recorded. Renowned throughout the northeastern provinces, he had no other healers in his family. He was neither mystical nor spiritual in his healing; he was a true empiric, having learned his art through trial and error. He was nearly illiterate and had never read any indigenous or foreign medical texts. Not visibly afflicted by leprosy at the time Menaut found him, Pen claimed to have healed himself of the disease. Although he had only vague notions of medicine, and knew nothing of treating other common illnesses, he had, in Menaut's estimation, almost perfect biological knowledge of leprosy.[93] He had settled in the area with several patients, or followers, some time before Menaut discovered the community. Village life in Kel Chey did not seem greatly different from life in other Khmer villages. If anything, the leper community's mutual affliction made it more cohesive than neighboring hamlets. According to colonial observers, the inhabitants had considerable faith in Pen's healing abilities and seemed very respectful and affectionate with each other. Indigents without money arriving in Pen's village worked with him cultivating the land. Although the community was somewhat isolated, it shared in a loose network of trade with other regional villages.[94]

Over the next several years, Menaut continued to monitor Kel Chey. At his urging, the colonial administration paid to build a new infirmary hut. By 1913 the village was comparatively prosperous.[95] Until 1915, the only interference by the French administration was the infirmary construction and a dossier on each leper in the village, kept by Menaut on his regular, somewhat informal visits. As mentioned above, the GGI had been pressuring the *résident superieur* to open a leprosarium in Cambodia for several years. Because Kel Chey functioned smoothly, its inhabitants little bothered by Menaut's occasional visits, the RSC decided that it would be ideal as the first official leprosarium of Cambodia. On June 7, 1915, the GGI decreed Pen "the indigenous director of the Protectorate Leprosarium" (directeur indigène de la Léproserie du Protectorat).[96] This, in the words of Menaut, "spelled [Pen's] ruin."[97]

The village of Kel Chey was renamed Troeung Leprosarium by the decree of June 7, 1915. It was not to be physically altered; it would continue to be run like an "ordinary Cambodian village," except for a few regulatory changes. Among these, Pen was now officially the *mekhum* (administrative village chief); inhabitants could not come or go without the permission of the sanitary authorities; and healthy individuals could only visit with permission and could not stay overnight within the village.[98] As assistance, the government would provide each villager a blanket and enough material for two outfits, as well as lighting oil and soap every three months. One of the most severe new rules decreed that any child newborn to a leper was to be removed within forty-eight hours and, when strong enough, sent to either healthy relatives or a charitable establishment (usually the Catholic Church).[99] Dossiers were to be kept on all lepers in the community, as well as these children.[100] In the following year, the RSC revised and intensified these regulations. No object "belonging to the leprosarium" could be carried off of the premises. Further, alcoholic beverages were strictly forbidden. *Internal* commerce was allowed, although commerce with the outside world was still forbidden. This bureaucratic distinction between interior and exterior would continue to recur. However, with neither effective surveillance by the authorities nor cognitive acceptance of such a distinction by the village inhabitants, official ostracism would in practice encourage local indifference.

The administration began providing daily meals to the villagers, which were also strictly rationed.[101] This act of magnanimity was made necessary by the concept of enforced isolation of the group. On some level the administration recognized the need for contact with the outside world to make a viable community as well as to provide the villagers with social purpose. Thus, regulations specifically allowed for the continuation of handicraft work and farming, although the products of these efforts "could not become the object of commerce with the exterior."[102] However, this dissuasion from external trade quite naturally led to disinclination towards generating the objects of trade. Idleness and indifference became noticeable among the inhabitants.[103]

Most severe among regulatory infractions was escape, which warranted an eight-day confinement.[104] Departure, however, would be a more accurate term than escape, since no physical barrier existed to free egress. To create an inexpensive physical barrier, the RSC actually toyed with the idea of relocating the leper village to an island in the middle of the Mekong River; the idea stalled for lack of funding.[105] Ultimately, Troeung was, in the words of one doctor, a "moral" rather than a "real enclosure."[106] Since inhabitants and some doctors doubted the disease's contagiousness, the threat of punitive measures against oneself, and not the moral threat of contagion to one's peers, was the only real deterrent to evasion.

## Citizenship and Anarchy

Within its problematic status as a permeable enclosure, Troeung interwove practices of governance with practices of science. The French were not alone in creating such leper villages in their colonies. During this time the British in India, the Americans in the neighboring Philippines, and colonies around the world were rounding up and confining lepers under increasing international scientific consensus on the properness of isolation.[107] In each case, the rationale for confining lepers consisted not solely of the danger they posed to the healthy population, but also of the benefits they could more efficiently receive within a monitored community. One scholar of colonial medicine has described the Culion leper colony in the Philippines as "an exemplary part of the colonial process of modern subject formation." The leper village was "a controlled laboratory of subject repositioning . . . [that] allowed colonial officials to think of 'citizenship' in terms of universalism and progress . . . ."[108] Examining the administrative rhetoric and bureaucratic regulations of Culion's operation, this scholar locates "the rituals of modern citizenship inflected in the protocols of the isolated disease community."[109] We can certainly trace similar administrative rhetoric in the myriad decrees and justifications for Troeung in colonial Cambodia.[110]

Such a framework is useful in understanding the mindset of metropolitan bureaucrats originating the policies about—or against—the lepers of French Indochina. Yet it is doubtful that the lepers perceived such "rituals of modern citizenship" as anything resembling their own citizenship formation, particularly since the "controlled" environment of the leper village was actually not controlled. Thus, in what ways were individuals operating within this rhetorical grid of transformation actually transformed? These conceptual changes could not occur without some sort of material change. Did the agents (both French and Khmer) who were assigned to effect these material changes for the sake of "subject repositioning" accept or even attempt their tasks?

Initially, the AM did not increase medical visits to Troeung. The colonial government built no walls and posted no additional surveillance or medical staff. The village was not physically altered. Nonetheless, the village of Kel Chey was conceptually transformed into Troeung Leprosarium. The villagers became, on paper, responsible self-policing citizens. Pen himself was transformed from the chosen leader of those seeking sanctuary into the administrative head of a medical confinement facility. He continued to act, however, as he had in his previous role. Confusion over his traditional authority in relation to his new and presumably higher authority was quick to come. In June 1917 the Prey Veng authorities detained a leper named Ek after an indigenous regional governor brought him to the attention of the medical chief. Although Ek had in his possession a note in Khmer signed by Pen permitting him leave from Troeung, the AM doctor promptly issued a certificate ordering Ek's internment in Troeung, "from where he probably escaped."[111] Ek, as a voluntary intern, had asked Pen for permission to return home briefly, both to check on his family and to procure some money. According to Ek's declaration to the authorities, he had verbally promised Pen he would return to Troeung quickly, but upon returning to his hometown he found his affairs in disarray. His wife was in a difficult financial situation; the *balaat* (an indigenous official)[112] had stolen much of the family property. The effort to recover his stolen property from the authorities, as well as his inability to earn money in short-term work, had delayed his promised return.[113] In his own statement to French authorities, the *mekhum* (a local administrator)[114] claimed that he and the *balaat* had recently searched Ek's property for money or valuables, apparently as reimbursement for some outstanding taxes that Ek owed, and could find none. Further, the Khmer officials had not found a house on the property (Khmer traditional homes are easily disassembled, moved, and reassembled). In his statement to the French authorities, the *mekhum* did not directly address Ek's accusation of theft. However, Ek's detainment was doubtlessly linked to his disputes with local authorities. His affliction made him vulnerable to the use of the colonial apparatus against him by local elites. Pen's note, handwritten in Khmer, had no legitimacy within this apparatus. Indeed, it is doubtful that the AM doctor bothered to have it translated. Ek was forcibly returned to Troeung in July 1917.[115]

This, and doubtless other similar episodes, eroded Pen's support among his patients. The offhanded contempt of colonial observers, the myriad rules he did not desire to enforce, and the loss of traditional authority soured Pen to this imposed position. Having developed a vehement hatred for the French colonial government very soon after Troeung's creation, Pen left the village he had founded soon after Ek was forcibly returned. He died alone in the forest of cholera on June 23, 1919.[116] Troeung continued to operate under the aegis of one of his students. However, the village had fallen into steady decline. The year after Pen's death, the health director visited Troeung. At that time the village had fifty-four inhabitants and approximately twenty huts, only three of which seemed inhabitable. The others were "falling apart" from the last rainy season, and the entire village was in a state of dilapidation.[117]

Through the 1920s, the AM issued a minimum of ten new internment decrees annually.[118] Although internment and reinternment papers greatly outnumbered release documentation, the population of the village did not increase correspondingly. Rather, from 1915 to 1929 it fluctuated irregularly between twenty-five and sixty.[119] Clearly, retention was a problem. In the early 1930s the number of inhabitants climbed rapidly. By 1932, 150 lepers were living at Troeung. Despite an ordinance forbidding unauthorized building within the village limits, a haphazard array of straw huts had sprung up. The entire effect was one of "insalubrity undignified for a medical establishment."[120] However, this was to be expected, due to social and financial neglect by the RSC. Despite hundreds of internment decrees, the administration had built only two new wooden pavilions in sixteen years. These two buildings, designed for approximately fifteen occupants each, were woefully insufficient for the population.[121] By 1935 the leprosarium still had no officially sanctioned new buildings, and nearly two hundred occupants (figure 6.2).[122]

Why more lepers chose to remain in Troeung and build their homes there in 1930 and 1931 is not entirely clear. The administration did not increase security, improve living conditions, or issue more internment decrees

Figure 6.2. Troeung Leprosarium, circa 1930. Courtesy of the National Archives of Cambodia.

annually. However, the global depression that began around this time may have made the small allowance provided to inhabitants by the colonial government increasingly attractive. Although the village grew larger in the 1930s, actual isolation within it did not, as various scandals would reveal. The AM chief would observe in 1932 that most villagers came and went at will. Others simply left without returning. The doctor complained, "It is rare that the native remains at the leprosarium until completely healed. As soon as his state seems returned somewhat to normal, he leaves—with force if you attempt to impede him—and foregoes all medical surveillance."[123]

By the mid 1930s, Troeung had developed a notorious reputation in the province. On March 7, 1934, the village was the site of a violent public lynching of one of the most unpopular lepers. Although the event was reported to the administration, there is no record of an investigation.[124] However, in 1935 the reports of a Troeung-based counterfeiting ring spurred an extensive investigation by the colonial government. This investigation was motivated not only by financial or juridical concerns; fear of contagion-carrying money may also have played a part.[125] All of the lepers questioned by the authorities were married, although their wives and children did not live with them in the village. These men traveled freely, often going to visit their families, with whom they maintained social ties and some financial responsibility. Some had been interned at Troeung forcibly, and others lived there voluntarily. Regardless of how they had arrived there, they frequently traveled to nearby towns, where they were passing the fake money in question. With the exception of one purchase of lead shot (presumably to produce more counterfeit coin), their purchases were largely innocuous. One counterfeiter admitted going to a nearby village to buy chickens and "edible products." Several others purchased products for making rice alcohol. The investigators concluded that two lepers were responsible for producing the counterfeit coins, but that almost everyone in the leprosarium was freely using them. There were, however, no extravagant purchases; villagers were spending the money almost exclusively on foodstuffs. The Kompong Cham *résident* would resignedly observe, "No local disciplinary authority exists at Troeung, and . . . being unable to intern the culpable parties in the Kompong Cham prison, I find myself obliged to leave this entire sad world at liberty."[126]

Registering the low morale and morality in 1934, as well as the constant evasions at Troeung in this year, the Kompong Cham medical chief lamented, "I believe . . . that the best way of keeping lepers in the leprosarium consists, not in imprisoning them, but in placing them in living conditions that are as normal as possible, that is to say, to give them the same way of life in the leprosarium as they previously lived in their village of birth. This goal is perfectly realizable if one has the will."[127] This is indeed an ironic observation for a village that, nineteen years earlier, had been very much like "their village of birth." In many ways the disastrous consequences had been predictable. If we return to the administrative discussions immediately before Troeung's 1915 creation, we find that the Kompong Cham medical chief had voiced no objections to creating a leprosarium at Kel Chey village, but did predict that rigorous rules would make such an effort a failure. The success of Kel Chey had been in its position as a voluntary sanctuary. The Kompong Cham *résident* agreed with the doctor's sentiments, noting that Kru Pen should be maintained if possible, due to his central role within the community.[128] Forced isolation was known to be widely unpopular.[129] After Troeung was decreed and the various other leprosy regulations were enacted for the territory, medical staff observed that the drastic decrease in "this category of diseased" coming for hospital consultations was directly due to the constant increase of "entirely new regulations limiting . . . individual freedom."[130] Numerous other doctors periodically commented through the 1920s and 1930s that fear of internment at Troeung made lepers hide from the authorities rather than come for treatment.[131]

Finally, the leprosarium was costly, and for this reason it was subverted by the very officials whose cooperation it required. The health services continually struggled to budget for the upkeep of Troeung lepers. With the global depression of the 1930s, this budget was foisted from one unwilling party to another. The cost was initially transferred from the country's general budget to the provincial budget of Kompong Cham, where Troeung was located. By 1935 each leper's maintenance was charged to the budget of his province of origin.[132] As these changes were made, local AM chiefs were more inclined, perhaps pressured, to recommend isolation at home rather than in the leprosarium. This was certainly the opinion of the health director in 1937, who grumbled that provincial officials turned a blind eye to lepers rather than reporting them and incurring their Troeung expense.[133]

Troeung was clearly not an institution of order or healing. The internees did not conceive of themselves as model citizens; administrators expended little effort in encouraging such a transformation. Authority and discipline—of either the traditional, bureaucratic, or technological variety—did not exist at Troeung. Rather, the inhabitants were placed under what to them were senseless rules created for a foreign purpose, and then largely forgotten. The various regulations, their sporadic enforcement, the occasional medical surveillance, the minimal handouts, and so forth served to create and enforce a sense not of citizenship, but of alienation. Despite the official rhetoric presented to metropolitan authorities and the international network of public health officials, Troeung was a failure as an exercise in modernity. If anything, its regulation encouraged anarchy and indifference.[134]

The leprosarium continued to experience scandals and overcrowding through the 1930s and 1940s. Troeung ultimately could not be undecreed, nor were conditions successfully ameliorated in later years, although a few efforts were made.[135] Menaut, who had befriended Pen so many years earlier, would bemoan of the entire endeavor (perhaps with some guilt over his own role in the affair):

> The clumsy and unjust intervention of our administration quickly caused the destruction of the eminently laudable efforts of this man [Pen], . . . [and] ruined this private Colony which only asked to live, in order to replace it by some official thing, in order to regulate it[;] the most that one can say of this effort was that it forced the Protectorate to spend excessively to end up in most pitiful failure and to plunge Pen back into the primitive's state of mistrustful savagery.[136]

Still, the RSC did take steps to assure that the same mistakes were not repeated. Two other private leper villages were reported to the RSC in the 1920s and the 1930s, both run by *kru khmer*. The AM staff monitored one of the private leper villages in Siem Reap province for several years. When the Siem Reap *résident* suggested turning the village into the province's official leprosarium in 1933, the health director advised the RSC that it remain private. The RSC agreed.[137]

## Chimera in a Cage

The story of "Cambodian colonial leprosy" is the story of neither leprosy nor Hansen's disease nor *chomngu khlong*. Although these different identities were constructed in negotiation with each other, the various constructs did not efface each other, nor did they effectively hybridize. Multiple social definitions coexisted with little interaction—a sort of plural society of epistemologies. If anything, the leper in Cambodia was a disease chimera, an uneven and senseless mix of several constructs hobbled together. Policies constructed around this chimera were ultimately ineffective.

In one of the most exhaustive surveys of leprosy conducted in Cambodia, Dr. Menaut generously estimated that in 1915 one leper existed for every 1,700 inhabitants. Despite his desire to draw attention to what he considered the problem of leprosy in the country, he noted that the number of lepers did not seem to have grown appreciably since the first detailed surveys conducted a decade earlier.[138] Fifteen years later, a career doctor in Cambodia would write, "Personally, my impression after 20 years of practicing in Indochina is that leprosy is mostly in regression in the country."[139] Forty years after the alarm was raised, the epidemic did not come, despite the lack of effective prevention or treatment. By World War II, the disease no longer created a sense of regulatory urgency among public health officials. Lepers still could not be cured, but neither were they the seeds of an impending epidemic of deformity and death. An effective treatment for leprosy was not found until the birth of the synthetic drug industry and the availability of chemotherapeutic drugs in Cambodia after World War II.[140] However, isolation as a government policy, even when the justification was gone, would continue on its own inertia. In Cambodia, leprosy incarceration would be a policy of the postcolonial government well into the 1960s.[141] Even though the myth of the chimera was dispelled, the imaginary creature continued to be caged.

SEVEN

# Cultural Insolubilities

World War II changed the global geopolitical map. Vichy's appointment of Jean Decoux as governor general of Indochina in August 1940 marked a period of tenuous French hold on the colony. During the war, many AM programs quietly disappeared. With the Japanese Operation Bright Moon on March 9, 1945, and Japanese declaration of Cambodian independence shortly thereafter, French colonialism in Cambodia effectively ended, despite the return of Allied control six short months later.[1] Although France—exceptional among the imperial powers—was slow to accept it, the age of high colonialism was over.[2]

## Transformations

The AM effectively functioned in Cambodia from 1907 to 1940. Immediately after World War II, colonial medicine limped on after a fashion. In Phnom Penh, the Mixed Hospital was renamed Preah Ket Melea Hospital, but remained largely French-staffed. A full-fledged Pasteur Institute, also headed by French Pasteurians, opened in Phnom Penh in 1953.[3] Other institutional forms of French medicine also continued. Many medical programs instituted in the colonial period continued after the acts on which they were based were scientifically outmoded. For instance, Cambodian researchers continued to experiment on the nativization of quinquinas trees well after synthetic drugs were developed to treat malaria.[4] The independent Cambodian government still required sanitary passports after these passports were no longer used in Europe.[5] Troeung Leprosarium continued to receive lepers until the 1970s. The most obvious medical legacy of French colonialism is in the buildings and institutions that still stand today. Western medical services are still provided in the now-dilapidated district hospitals originally built by the colonial government. Further, after independence, King Norodom Sihanouk presided over the opening of dozens of new rural infirmaries built on the AM model. After the pomp of royal inauguration ceremonies ended, however, these infirmaries often served no medical purpose, as the government would move necessary equipment to the next clinic inauguration ceremony. The building itself as a symbol—the form without its function—seemed sufficient to represent progress.

This phenomenon echoes the idea of a Southeast Asian theater state—the role of theatricality and the presentation of the form—in creating a legitimate state authority in the Khmer context.[6] The nature of state authority and responses to it involved both a traditional and a newly instituted form of polity. The preexisting polity, we may speculate, relied heavily on the staging of power. However, throughout the colonial period the French government was also "staging" its own power. Science and medicine formed part of this theater for, as historian Gyan Prakash has observed for British India, staged colonial representations of science were designed to produce "awe, magic, and marvel" in the eye of the indigenous beholder.[7] This staging of power not only involved the overt display of miracle cures, but also entailed creating a myth of the infallibility of Western medicine, as we have seen in chapter 2. However, the myth's creators were not solely concerned with colonial power. They were also concerned with legitimizing the professional autonomy of those within the medical and scientific research fields. Their domain of medical knowledge had to be both distinct from and authoritatively superior to that of the lay population (European and indigenous). Yet constituents had to be educated *and* impressed. Thus, patients needed to be informed or at the very least awed enough to cooperate with medical measures, yet uninformed enough to accept the authority of this specialized knowledge. As in the metropole, specialists were performing a balancing act of both including and distancing the knowledge base of the general population. Specific to the colony, the domination of the colonial state and the cultural foreignness of the doctor complicated the issue. The staging of scientific knowledge as political power, and its imitations of the staging of indigenous political authority, was a symptom of the multiple layers of domination and negotiation at work in the colonial domain.

What of the impact of the Cambodian experience on French medical research? In French overseas scientific endeavors, medical research grew in importance in the late nineteenth century as the political value of medicine increased.[8] Because diseases of the tropics presented a significant challenge to efficient colonial exploitation, they served as a tremendous impetus to research. These microbial challenges of colonialism partly drove the revolution in germ theory.[9] Further, imperial competition buttressed the international race to discover the causes and cures of diseases from yellow fever to malaria. Particularly for Indochina, research on the plague and cholera was imperative. The ecology of the tropics was clearly different from that of temperate climes, and it provided different opportunities for research.[10] Indeed, the conditions of colonialism led to the growth of tropical medicine as a separate disciplinary field.[11]

Thus, colonialism clearly had a profound impact on the structure of scientific and medical research. However, we cannot point to a medical development, with the partial exception of plague research, that was specifically dependent on France's relationship with Indochina or Cambodia. Neither can we point to a specific therapy "learned from" Cambodia. However, to find such "discoveries" is of little value to the purpose of this study. This may seem strange, as our focus has been transformations through the medical exchange in Cambodia. However, a central claim of this study is that these transformations are not to be found in simple exchanges of techniques or technologies.

## Demographic Impact

Since the French medical service made little headway in changing Khmer interpretations of sickness and death, the impact of French medicine may perhaps be better found in morbidity and mortality statistics than in transformations in cultural meanings surrounding death and disease. The population of Cambodia quadrupled between 1863 and 1954 (table 7.1). Imposition of Western medicine would thus seem to correlate strongly with increased survivorship. However, several factors discount, or at the very least deemphasize, the role of French medicine in this trend. Although we have little reliable data to analyze fine details of demographic trends (immigration, emigration, birth, death, life expectancy) during the first half of the twentieth century, we can make some generalizations from the record. The French colonial period saw a large influx of immigrants into Cambodia, predominantly from Vietnam and China. However, immigration alone did not account for the quadrupling of Cambodia's population. Thus, we may speculate that more births were occurring. Available data on maternalism, infant care (see chapter 5), and vaccination (see chapter 4) indicate that infant mortality was little affected by Western medicine. Considering that decrease in infant mortality is one of the major reasons for the initial rise in life expectancy in countries going through a demographic transition, it is doubtful that life expectancy was significantly affected by medical intervention on

**Table 7.1. Population of Cambodia**

| Year | Estimated population |
|---|---|
| 1875 | 1,100,000 |
| 1900 | 1,500,000 |
| 1921 | 2,520,000 |
| 1931 | 2,806,000 |
| 1939 | 2,195,000 |
| 1948 | 3,748,000 |

Sources: *Annuaire statistique de l'Indochine* and Jean Delvert, *Le paysan cambodgien* (Paris: Mouton, 1961).

children during this period. Further, the spotty record of disease morbidity and mortality tallies available in annual medical reports indicates that the number of deaths from preventable diseases was not notably affected (see chapters 3 and 4). While the colonial government did develop programs that significantly decreased the incidence of smallpox and yaws, these were small gains that could not account for such a tremendous population increase in such a short period. Thus, we may speculate that a major factor in population growth is increased births (not survivorship after birth), fueled in large part by immigration and by decreased death and instability from conflict (as distinct from infectious disease).[12]

Is political stability the only factor driving this growth? Studies of modern medicine in Europe proffer several alternative antecedents to the rise in population. Thomas McKeown, for instance, famously argued that improved nutrition, irrespective of medical developments, generated the tremendous population boom in nineteenth-century Europe.[13] In Cambodia, the majority of the population remained poor and undernourished during this time, although a fortunate sliver of the population found increased prosperity from colonial economic opportunities.[14] No nutritional or supplementation programs existed in colonial Cambodia. Considering these facts, nutrition was likely unimportant in the observed population increase. Since McKeown's original study, other scholars have asserted that wider-ranging social factors, including education, city planning, and changed behaviors, are more significant determinants of population growth in a given community.[15]

Attribution of population growth to colonial medicine becomes more untenable when we consider that diseases with the greatest mortal impact, such as malaria and dysentery, were little affected. Further, Western medical practices did not infiltrate a significant distance beyond the few urban centers in the country. Outside those urban areas, the French colonial government did not effectively change behaviors, implement disease prevention campaigns, improve nutrition, ensure clean water, provide garbage removal, or educate the population. However, it did end centuries of civil war and permit an extended period of relative social stability.[16] Southeast Asia between the sixteenth and eighteenth centuries had a relatively sparse population due to insecurity, raiding, and warfare. Cambodia immediately before French intervention had been decimated by decades of civil war. Historical patterns from the sixteenth to the eighteenth centuries reveal that populations rose *rapidly* "whenever conditions of stability were assured."[17] It seems most reasonable to speculate that this historical pattern also applied to twentieth-century Cambodia.[18]

## Pluralities

John Sydenham Furnivall famously wrote that colonialism in Southeast Asia tended to create a "plural society . . . comprising two or more elements or social orders which live side by side, yet without mingling, in one political unit." These societies meaningfully commingled only in the marketplace.[19] Like Furnivall's idealized plural societies, diseases in the colonial context had their own pronounced social plurality. One disease paradigmatic of this plurality was leprosy, as discussed in chapter 6. The affliction and its associated sets of meanings set it apart from the Khmer term *chomngu khlong*. In Europe in the late nineteenth century, the medical community began to use the term Hansen's disease in preference to leprosy to divorce the *bacterial* condition from its social stigma. However, at the same historical moment, the French colonial government in many ways generated a uniquely "European" social stigma to justify confinement policies in the colonies. Leprosy, Hansen's Disease, and *chomgnu khlong* did not simply signify a disease; the choice of term used determined how the afflicted would be perceived and treated. The same difficulties could be seen in the pairings of terminology for diseases such as the plague/*chomgnu puh* and malaria/*kronh chanh*. Even basic terms such as "contagion" and "germ" had multiple significations. The technical advantage and programmatic goals of the colonial government in its efforts to enforce its own definitions set them apart from other, differing conceptions of disease. These definitions were further confounded by the growing gap in French between the specialized language of scientific medicine and the quotidian language for health and illness. In large part, the gulf in conceptualizations of illness between colonizer and colonized grew immensely with the development of microbiology and revolution in germ theory in the West.

If we use the analogy, perhaps apropos, of "mandalas" of medical influence, both French and indigenous medicine could operate in a single sphere.[20] Patients gave necessary tribute to both at times, but could fail to recognize either as permanent or supreme. In partial contrast, in contemporary Europe and America, science commands authority, as does medicine. However, the authority of the doctor is a relatively new construct of the late nineteenth century. This authority—whether constructed, as medical historian Paul Starr argues, on economic controls and monopolies of related systems to obtain legitimacy and dependence, or, as Charles E. Rosenberg argues, in conjugation with the power of ideas, the allure of innovation, the therapeutic promise of hospital-based technologies, and social changes—does not materialize in the Khmer context.[21] This is in large part because the social changes that drive these transformations do not occur in Cambodia. If we examine the economic premise for the authority of biomedicine, it assumes a provider monopoly, doctor-patient contractualism, and a capitalist economy. The Western doctor has a professional social distance from his patient. The patient does not need personal knowledge of the doctor, because the doctor's professional network enforces his legitimacy and generates his authority. However, the doctor never had such a monopoly in the Cambodian context. A village doctor could not just serve in an impersonal contractual role, relying solely on his medical talents for income. Potential patients did not have money to participate in the contractual cash economy demanded by modern medicine. In contrast, indigenous healers did not rely solely on their medical talents for survival; they were strongly integrated in the community as farmers, monks, housewives, and so forth. For these practitioners, medicine was not so much a profession as one of a variety of social roles they had to fulfill within their villages.

If we assume that medical authority is constructed through the power of ideas, we encounter the problem of incommensurability. Concepts of disease, causes, preventatives, and cures are bundled with broader concepts of the individual, the political structure, religion, and myriad other cultural constructs. The strength of a healing system depends not only on "objective successes; equally important for its continuing acceptance and support is the anchoring of its medical notions in the world view and, especially, in the sociotheoretical concepts of a population."[22] The colonial government could not transfer piecemeal the concept of scientific knowledge and authority without effecting profound epistemological changes in a variety of fields of knowledge. Indigenous societies did not accept many of the constructs imposed by the medical services; this was not due to political resistance but because these constructs were neither intellectually satisfying nor sensible in changed social contexts. Ultimately, the French could not produce allure for certain ideas, create prestige in medicine, or lure the brightest and best into a biomedical calling. In fact, many of the indigenous men who chose to enter the medical profession seemed to rely upon a personalized patron-client notion of authority. Ironically, during the Khmer Rouge period, the excesses of power were conflated with the excesses of Western knowledge, and the doctors of the country experienced a state backlash against a sort of authority that in truth did not exist.

Cambodia still has a multiplicity of thriving medical traditions. Most births and deaths still occur in the home. The overwhelming majority of Khmers turn to traditional medicine before resorting to Western medicine or visiting the hospital. In contrast to new developments such as karaoke, television, and cell phones—all seen as modern while not necessarily foreign—Western medicine is viewed as foreign while not necessarily modern.[23] This may be due to the lack of exposure Khmers have to many biomedical practices. In other words: biomedicine's perceived foreignness, in contrast to the cell phone for instance, is not due to its origin but its unfamiliarity. Today, as in 1930, country dwellers still give distance (over cost or the provider's competence) as the main factor determining choice of medical care.[24] However, biomedicine slowly gains proponents in the country. Traditional medicine, in recent years, has been forced to organize itself institutionally to gain validity in the health market. This is perhaps due both to the current Khmer project of nation building and to the Vietnamese occupation of the 1980s. The Vietnam government also has a traditional medicine association imbricated in the nationalist project.[25]

While the Western doctor's authority never materialized, the power of foreign medicine is today immense. Western pharmaceuticals are still very popular, unregulated, and widely available.[26] Khmers show little hesitation in ingesting pills in quantities and combinations that would certainly not be advised by the pharmaceutical producer or a Western medical doctor. Taken in such a way, they objectively do perhaps as much harm as good. Yet "untrained" drug sellers continue to peddle antibiotics, cough medicines, and painkillers as combined therapies for ailments both mild and severe, and patients continue to eagerly purchase such advice and treatment. This phenomenon is common in many postcolonial nations.[27] The most plausible explanation for this phenomenon is medicine's easy commodification and transportability. In other words, as a "thing," a pharmaceutical product is easily separated from the "regimes of value" that created it and imbued it with another social meaning.[28] The fungibility of the cultural meaning of material objects has increasingly become the focus of scholarly analysis in science studies and colonial studies, but is treated distinctly in those two disciplines.[29] However, as the anthropological research on pharmaceuticals and the analysis in this book reveals, the theoretical implications within each type of analysis can be combined to yield a fruitful framework for understanding scientific transformations under colonial control.

## Metanarratives of the Colonial Period

Historians can write the colonial experience as negotiation, debate, friction, or conflict between colonizer and colonized. All of these processes are in some sense dialectical, and as with all dialectical processes, colonialism eventually establishes some sort of cultural transformation (modernity, nationalism, capitalism) through synthesis (even if it is the historian rather than the historical actors that perform the task). Whether the metanarrative is framed from above (political histories), below (subaltern studies), or within (cultural studies), it can fail to acknowledge that the historical actors may not experience these transformations in those terms. Further, histories of medicine in the colonial context often present colonialism at the fore, as the engine of the narrative. In contrast, medical anthropologists and sociologists who study the medical practices of these same cultures in the postcolonial age often focus on the fit between imported and preexisting medical beliefs, practices, and experiences.[30] Histories of medicine in Europe and America also recognize the importance of wider social frameworks in determining both how medical practices change and how people accept medical concepts.[31] These considerations are dwarfed in medical histories of the colonial period, in favor of explanations based on the dynamics of colonization: coercion, resistance, control, surveillance, evasion, and so forth. Perhaps the term "colonial medicine" itself entraps the historian into thinking within this framework. Certainly the dynamics of colonization are important in how medicine was perceived; however, other oft-neglected factors also strongly influenced the colonial interaction. This study has approached medicine in a way that productive histories of medicine in "noncolonial" settings do: by examining in both descriptive and theoretical depth the wider cultural framework of medical ideologies and practice.

Culture is a problem for many disciplines, including history. Even the field of cultural history leaves many assumptions about culture untouched. No matter how ambitious we are, we cannot understand, grasp, or define entire cultural systems. Although this book has analyzed languages, institutions, religions, values, traditions, rituals, economics, politics, and science, it still has only remotely touched upon some of the aspects of Khmer and

French culture that affected the transformation of Western medicine in early-twentieth-century Cambodia. Our tableau is still largely incomplete. The framework of cultural insolubility gives some weight to the offstage processes we cannot recapture or emplot within our narrative. It also attends to the real phenomenon of cultural inertia and acknowledges, in some way, the right and rightness of the many Cambodians of the colonial period to stay at the margins of this Western, linear history. What they were doing was meaningful, even if opaque to historical documentation. The failures of French medicine were in part a result of the actions and choices of this silent majority. However, insolubility was not the only process at work, although it has been neglected in many historical studies. Cultures were evolving as well. Culture is adaptive, but not in the way that certain kinds of systems thinking allow.

It is instructive to examine current rhetoric in the field of public health in what are now postcolonial countries. For the historian examining these contemporary issues, it is quite surprising how little the discourse on culture and modernization has changed. Social scientists and development workers model, deconstruct, segment, and analyze culture as discrete behaviors or domains that can either facilitate or obstruct planned interventions. Many of these interventions are eerily similar to colonial programs. Take, for example, the public health practice of "ring vaccinations," a standard protocol used today by the Centers for Disease Control in the United States and overseas to contain epidemic outbreaks. With the ring vaccination approach, "patients with suspected or confirmed smallpox are isolated, and contacts are traced, vaccinated, and kept under close surveillance, as are the household contacts of those contacts."[32] Ring vaccinations have been discussed as applicable across a broad range of new epidemiological threats, such as avian or H1N1 flu and bioterrorism. Epidemiologist and public health leaders view this method as an effective and *just* strategy to contain epidemics, even though it regularly entails coercive force against "noncompliers." Consider how, during a smallpox epidemic a century ago in Cambodia, the municipal doctor of Battambang complained, "As soon as an epidemic breaks out, the inhabitants avoid coming to alert the authorities because, more than the disease, they fear our methods of prevention and disinfection, which they consider troublesome . . . . The population carefully hides its sick, furtively buries its dead."[33] The colonial government addressed the "problem" by trying to increase regressive measures and provide more policing powers to the health services. Historians look upon such episodes with an enlightened disapprobation, even as today the same techniques continue with the self-congratulatory approval of public health researchers.

Epidemiologists and public health policy makers continue to find ways to minimize the very real problem of individual resistance to coercive techniques in the name of the greater good. Top-down policies are enacted for the welfare of the population; individual will and cultural sensibilities are secondary. For more mundane issues, today just as in the colonial period, health workers are urged to include (co-opt) local elites (now called community leaders) in program planning.[34] These programs are modeled from above but involve local "participation," which often means little more than facilitation.[35] The language is new, but the treatment is decades old. Just as the French colonial doctor promoted health interventions that took into account the untrained thinking of the rice farmer by "drawing from his observations of daily life and reasoning with him by analogy and image,"[36] today's public health worker may encourage HIV "prevention programs that incorporate social cognitions" of workers in the developing world.[37] In this fashion, development workers, including public health workers, can feel that their clients' cultures have been "taken into account" and "respected." While the modern and colonial public health worker's interventionist stance compels some of this instrumentalist and reductionist thinking, historiographical treatment of culture is also not sinless. Our analysis of the relation between culture and modernization in the colony also reduces it by assuming that the individuals within the culture are deliberately making choices to protect both their social identities and the coherence of their cultural system. Our sympathy is with the people; we give them agency since they comprise the cultural group. But the whole is more than the sum of the parts. What the stories in this book and the continuing problems in development work reveal is that we have little understanding of the mechanisms that help cultures keep their coherence. I am not giving cultures agency here; rather, I am proposing that cultural insolubility is somewhat independent of individual will or deliberate conscious acts of people within a cultural system. Thus, when we characterize people as defiant or resistant, we may be mischaracterizing them. Culture is process as much as structure and practice. Whether they know it or not, the attitudes, actions, and even thoughts of individuals are constrained by their lived cultures.

This study has approached the colonial domain with an awareness of the multiple perspectives operating on medical history, even (or especially) in the colony. Bringing the many characters' stories together in this way brings attention to the vast ontological gap between Khmer and French understandings of medicine, but also to the unexpected similarities in internal social dynamics within different cultural spheres. This is a history of men and women and how they have negotiated the multiplicity of their identities, ideas, and actions. But this is also a story of how norms, institutions, geography, economics, and various other factors directly and indirectly affected those negotiations. Ultimately, human beings dwell somewhere between the realm of ideas and practice. In suggesting parts of lived experience that the historical record cannot give us directly, as well as multiple narratives of how medical thinking was constituted and transformed among different groups in the colonial setting, this book is meant to be more than the sum of the individual stories it contains.

# NOTES

INTRODUCTION

1. See Nicholas Thomas, *Entangled Objects: Exchange, Material Culture, and Colonialism in the Pacific* (Cambridge, MA: Harvard University Press, 1991), Nelly Oudshoorn and T. J. Pinch, *How Users Matter: The Co-Construction of Users and Technologies*, Inside Technology (Cambridge, MA: MIT Press, 2003).
2. Clifford Geertz, *The Interpretation of Culture* (New York: Basic Books, 1973), p. 5.
3. Penny Edwards, *Cambodge: The Cultivation of a Nation, 1860–1945* (Honolulu: University of Hawaii Press, 2006), p. 243.
4. Anne Ruth Hansen, *How to Behave: Buddhism and Modernity in Colonial Cambodia, 1860–1930* (Honolulu: University of Hawaii Press, 2007), p. 13.
5. Max Weber, *The Theory of Social and Economic Organization* (New York: The Free Press, 1947).
6. Oudshoorn and Pinch, *How Users Matter: The Co-Construction of Users and Technologies*, chapters 2, 3, and 5.
7. Christopher Goscha, *Vietnam or Indochina? Contesting Concepts of Space in Vietnamese Nationalism, 1887–1954* (Copenhagen: NIAS Books, 1995).
8. See, most notably, Teresa Meade and Mark Walker, eds., *Science, Medicine, and Cultural Imperialism* (New York: St. Martin's Press, 1991). See also Roy MacLeod and Milton Lewis, eds., *Disease, Medicine, and Empire: Perspectives on Western Medicine and the Experience of European Expansion* (New York: Routledge, 1988), particularly the last three essays. For a fine example of theoretical sophistication within this model, see David Arnold, *Colonizing the Body: State Medicine and Epidemic Disease in Nineteenth-Century India* (Berkeley: University of California Press, 1993). For information on science more generally as tool of empire, see Daniel R. Headrick, *The Tools of Empire: Technology and European Imperialism in the Nineteenth Century* (New York: Oxford University Press, 1981), Daniel Headrick, *The Tentacles of Progress: Technology Transfer in the Age of Imperialism, 1850–1940* (New York: Oxford University Press, 1988).
9. For example, Meghan Vaughan's monograph on colonial medicine in Africa is an excellent study of the creation of the "African" through European medical discourse, but it provides little information on the actual experience of Africans. Megan Vaughan, *Curing Their Ills: Colonial Power and African Illness* (Stanford: Stanford University Press, 1991).

10. See Biswamoy Pati and Mark Harrison, eds., *Health, Medicine, and Empire: Perspectives on Colonial India* (Hyderabad, India: Orient Longman Ltd., 2001). Pati and Harrison note in the introduction that, "Even critics of the government's medical policy note that its limitations were due as much to indigenous indifference, or, even, hostility to medical and sanitary intervention, as to any weakness of official commitment." p. 4.
11. The two monographs in English on colonial medicine in Southeast Asia are by van Heteren and Manderson. A. de G. M. van Heteren, M. J. D. Poulissen, A. de Knecht-van Eekelen, and A. M. Luyendijk-Elshout, eds., *Dutch Medicine in the Malay Archipelago 1816–1942* (Amsterdam: Rodopi, 1989), Lenore Manderson, *Sickness and the State: Health and Illness in Colonial Malaya, 1870–1940* (Hong Kong: Cambridge University Press, 1996). For history of colonial science in Southeast Asia more generally, see the two monographs by Pyenson. Lewis Pyenson, *Empire of Reason: Exact Sciences in Indonesia, 1840–1940* (New York: E. J. Brill, 1989), and *Civilizing Mission: Exact Sciences and French Overseas Expansion, 1830–1940* (Baltimore: Johns Hopkins University Press, 1993).
12. Monnais-Rousselot's encyclopedic study of the Assistance Médicale in Indochina is explicitly anti-Headrick (see, for example, p. 32). Laurence Monnais-Rousselot, *Médecine et colonisation: L'aventure Indochinoise* (Paris: CNRS Editions, 1999). A more extreme effort to rehabilitate the "colonial doctor" can be found in Lapeysonnie, *La médecine coloniale: Mythes et réalités* (Paris: Seghers, 1988). For the demographically devastating effects of colonialism, see Ken De Bevoise, *Agents of Apocalypse: Epidemic Disease in the Colonial Philippines* (Princeton, NJ: Princeton University Press, 1995), Rita Headrick and Daniel R. Headrick, *Colonialism, Health and Illness in French Equatorial Africa, 1885–1935* (Atlanta, GA: African Studies Association Press, 1994). Both point to the colonial government's role in upsetting local human ecological systems, causing dislocation, epidemics, and death.

CHAPTER ONE

1. NAC RSC 32226. Letter from Mayor Paul Collard to RSC, July 10, 1907. The mayor claimed that he had instructed the health staff to keep a list of pig owners to indemnify; he had even intended to allow the owners themselves to set the pigs' values. It is doubtful, with the haste of the killings and the confusion that ensued, that these orders had been followed.
2. Zymotic theory, centuries old, argues that filth generates disease.
3. Paul Louis Simond, one of the early Pasteurians in Indochina, is credited with what are now seen as classic biological experiments linking the plague to flea bites. Published in 1898, his findings were at first roundly criticized. By the early twentieth century, however, fellow researchers accepted his hypothesis, with medical practitioners soon following suit.
4. "Une ridicule histoire de porcs incinérés." NAC RSC 12624. For a longer history of this incident, see Sokhieng Au, "Indigenous Politics, Public Health and the Cambodian Colonial State," *South East Asia Research* 14, no. 1 (2006).
5. French administrative rhetoric in Cambodia rarely distinguished Khmers from other ethnic groups (lumping them all under "*cambodgien*"), and sometimes did not distinguish Khmers from the Vietnamese population (using rather the general appellation "*indochinoise*"). Although the majority of the indigenous population was ethnic Khmer, there also existed a sizeable heterogeneous ethnic minority mix. For the purposes of this study, "Cambodian" refers to the population of Cambodia. Khmer

refers to the ethnicity. Annamite is the French term for ethnic Vietnamese, and will be retained when it refers to French classifications.

6. The Chams are an ethnic group on mainland Southeast Asia. They permanently lost the territorial basis of their kingdom (Champa) to Vietnamese expansion in the seventeenth and eighteenth centuries. They are now dispersed in parts of Vietnam, Cambodia, and Laos, where they maintain their own language, customs, and religion. French scholar Adhemard Leclère's history of Cambodia, particularly chapter 4, has a good chronology of this period. Adhemard Leclère, *Histoire du Cambodge depuis le premier siècle de notre ere* (Paris: Librarie Paul Geuthner, 1914). Li Tana's study of Nguyen Cochinchina describes some of the cultural and economic aspects of Vietnamese southward expansion through Cham territory. Li Tana, *Nguyen Cochinchina: Southern Vietnam in the Seventeenth and Eighteenth Centuries* (Ithaca, NY: Southeast Asia Program Publications, 1998).
7. However, to judge by existing folktales, most notably the popular story of Preah Ko (sacred cow), the anxiety about sovereignty existed at some level among the general population.
8. There are several problems in such an analysis. Conservatism does not necessarily represent stagnation. Also, the Khmer population is not the same entity as the Khmer polity. Although tribute may have been sent from as far away as Burma to the Angkorean Khmer king, this did not mean that the distribution of ethnic Khmers extended much further in the twelfth century than in the nineteenth century. Part of the apologeticist logic for French colonialism emphasizes this notion of inexorable decline, arguing that France stopped the disappearance of Cambodia—that if not for French interference, Cambodia as a sovereign territorial entity would have disappeared, and that "sovereignty" and "territory" were saved. This makes little sense if we consider that France essentially usurped the sovereignty of the Cambodian king and that French colonization thus represented the total, if temporary, loss of both territory and sovereignty. If we base the argument of decline solely on territorial loss, this again confronts the changing definition of polity. If the argument is premised on the decreasing size of a cultural and ethnic Khmer population, there is no indication that it was in decline. For example, Siam governed the two northwestern provinces of Siem Reap and Battambang from 1795 to 1907, but there is no strong evidence that the Siamese attempted to change Khmer cultural practices or eliminate the ethnic majority of the region. The fluid nature of political control between different strong centers, the waxing and waning of so-called Southeast Asian mandalas, was fundamentally altered and forced onto a linear historical time frame when European powers imposed geographical boundaries, the notion of hierarchies of societal evolution, and our specifically Western view of history as a progression through time.
9. For a succinct formulation of a contest state, see Michael Adas, "From Avoidance to Confrontation: Peasant Protest in Precolonial and Colonial Southeast Asia," *Comparative Studies of Society and History* 23, no. 2 (1981): p. 218.
10. Robert Heine-Geldern, "Conceptions of State and Kingship in Southeast Asia," *The Far Eastern Quarterly* 2, no. 1 (1942). The instability of dynastic lines is key to the mandala concept of kingdoms as well as to the formulation of the Southeast Asian "man of prowess." Wolters, for example, ties the man of prowess concept to an indifference towards lineage descent. O. W. Wolters, *History, Culture and Region in the Southeast Asian Perspectives* (Pasir Panjang, Singapore: Institute of Southeast Asian Studies, 1982), p. 31.

11. To facilitate a smooth succession, the king could declare an heir presumptive, the *obyureach*, during his lifetime. Further, the king could choose to relinquish his throne to the *obyureach*, taking the advisory position of *apyoreach*, to ensure the succession during his lifetime. Despite these safeguards, the transition often was not smooth. The title of *obyureach* did not ensure the bearer's acceptance by the royal court, as in fact the five great ministers of the kingdom, the Council of Ministers, ultimately had to approve the successor at the king's death. Milton E. Osborne, *The French Presence in Cochinchina: Rule and Response (1859–1905)* (Ithaca, NY: Cornell University Press, 1969). Later in the French colonial period, the number of ministers was decreased to four. See Réné Morizon, *Monographie du Cambodge*, Exposition Coloniale Internationale, Paris (Hanoi: Imprimerie d'Extrême-Orient, 1931).
12. For a review of the political trends among mainland Southeast Asian kingdoms during the early modern era, see Victor Lieberman, *Strange Parallels: Southeast Asia in Global Context, C 800–1830* (Cambridge: Cambridge University Press, 2003).
13. Father Gabriel Quiroga de San Antonio, *A Brief and Truthful Relation of Events in the Kingdom of Cambodia*, trans. Antoine Cabaton (Bangkok: White Lotus Press, 1998 [1604]). Although San Antonio's accounts are not highly reliable, the introduction by translator Cabaton provides a good overview of Spanish and Portuguese misadventures in Khmer civil wars of the sixteenth century. A somewhat more detailed analysis can be found in Bernard Philippe Groslier, *Angkor et le Cambodge au XVIe siècle d'après les sources portugaises et espagnoles* (Paris: Presses Universitaires de France, 1958).
14. Vietnam is a postwar name of unified Annam, Tonkin, and Cochinchina. For a history of the naming of Vietnam, see Christopher Goscha, *Vietnam or Indochina? Contesting Concepts of Space in Vietnamese Nationalism, 1887–1954* (Copenhagen: NIAS Books, 1995), introduction and part 1.
15. Historians differ on whether these rebellions were culturally or politically motivated. Osborne argues that Ming Mang's efforts at cultural transformation, particularly the persecution of the Buddhist sangha, were the impetus for this uprising. Osborne, *The French Presence*, chapter 1. David Chandler, on the other hand, finds that the growing rebellions in the region were due largely to political changes, specifically the replacement of Khmer indigenous governors with Vietnamese. This new policy encouraged indigenous leaders, whose political powers were threatened, to stir up their constituent populations against Vietnamese rule. David Chandler, *A History of Cambodia*, 3rd ed. (Boulder: Westview Press, 1996), chapter 7.
16. Louis de Carné, *Travels on the Mekong: Cambodia, Laos, and Yunnan*, 2 ed. (Bangkok: White Lotus Press, 2000 [1872]), p. 11. A foreign observer in the 1960s would blame the antagonism between the Vietnamese fishermen and the Khmer rice farmers on unidentified "painful memories of past centuries, transmitted orally, [that] seem to prevent their rapport." Gabrielle Martel, *Lovea, village des environs d'Angkor: Aspects démographiques, economiques et sociologiques du monde rural cambodgien dans la province de Siem-Reap* (Paris: École Française d'Extrême Orient, 1975), p. 43.
17. Duong was an uncharacteristically progressive Khmer monarch who attempted various political and economic reforms during his thirteen-year reign, including reforms to rules of succession. Governing a society depleted of economic resources and exhausted by constant civil war, he did not live long enough to effect real change.
18. John F. Cady, *The Roots of French Imperialism in Eastern Asia* (Ithaca, NY: Cornell University Press, 1954), chapter 1. For more detail on the complicated relationship between the French church and the military, see J. P. Daughton, *An Empire Divided:*

*Religion, Republicanism, and the Making of French Colonialism, 1880–1914* (Oxford: Oxford University Press, 2006).

19. C. M. Andrew and A. S. Kanya-Forstner, "The French Colonial Party: Its Composition, Aims and Influence," *Historical Journal* 14, no. 1 (1971), Henri Brunschwig, *French Colonialism 1871–1914: Myths and Realities* (New York: Praeger, 1964), Cady, *Roots of French Imperialism*.
20. For more details of this period see Osborne, *The French Presence*, chapter 1.
21. Ibid., chapter 10.
22. For a general review of French economic activity during this time, see John Tully, *Cambodia under the Tricolour: King Sisowath and the 'Mission Civilisatrice' 1904–1927* (Clayton, Victoria: Monash Asia Institute, 1996), or his follow-up book, *France on the Mekong: A History of the Protectorate in Cambodia, 1863–1953* (Lanham, MD: University of America Press, 2002).
23. Charles Robequain, *The Economic Development of French Indo-China* (New York: Oxford University Press, 1944), p. 9.
24. Cochinchina was again the exception.
25. Osborne, *The French Presence*, chapter 10.
26. Some scholars argue that the true colonial period was 1884–1954.
27. As mentioned earlier in the chapter, peace was a relatively new development of the second half of the nineteenth century. Chhuong Tauch, *Battambang during the Time of the Lord Governor* (Phnom Penh: CEDORECK, 1994), chapter 7. See also Evelyn Porée-Maspero, *Cérémonies privées des cambodgiens* (Phnom Penh: Éditions de L'Institut Bouddhique, 1958).
28. Osborne, *The French Presence*, chapter 1. Mahout would observe of one mountain village, "Quite alone and independent amidst their forests, they scarcely recognise any authority but that of the chief of the village, whose dignity is generally hereditary." Christopher Pym, ed., *Henri Mouhot's Diary: Travels in the Central Parts of Siam, Cambodia and Laos during the Years 1858–61*, Oxford in Asia Historical Reprints (Kuala Lumpur: Oxford University Press, 1966), p. 71.
29. The first Khmer-language novel was not published until 1938. Chandler, *A History of Cambodia*, p. 159.
30. Tauch, *Battambang during the Time of the Lord Governor*, chapter 7. Tauch describes a prominent poet. We also have an extensive analysis of an oral poem describing the Vietnamese interregnum, written in the late nineteenth century. Sok Khin, *L'annexion du Cambodge par les vietnamiens au XIX siècle* (Paris: Editions You-Feng, 2002). See also Solange Thierry, *Le Cambodge des contes* (Paris: Editions L'Harmattan, 1985).
31. Alain Forest, *Le culte des genies protecteurs au Cambodge: Analysis et traduction d'un corpus de textes sur les Neak Ta* (Paris: Harmattan, 1992), Introduction. Georges Condominas, from a study of Vietnamese highlanders, argued that all of Southeast Asia has a parallel significant distinction between the mountain people (wild) and the plain dwellers (rice growers/cultivators). G. Condominas, *L'espace social à propos de l'Asie du Sud-Est* (Paris: Flammarion, 1980). Ang Choulean recognizes a *prey/srok* opposition as well as a savage/civilized opposition. Ang Choulean, *Les êtres surnaturels dans la religion populaire khmère* (Paris: CEDORECK, 1986), p. 159.
32. Many travelers provide vivid accounts of animal attacks in the rural areas. See for example, Pym, ed., *Henri Mouhot's Diary*, p. 70. Also see Francis Garnier, *Travels in Cambodia and Part of Laos: The Mekong Exploration Commission Report (1866–1868)*, trans. Walter E. J. Tips, 2 vols., vol. 1 (Bangkok: White Lotus Press, 1996 [1885]).

33. R. Baradat, "Le samrê ou péâr: Population primitive de l'ouest du Cambodge," *BEFEO* XLI, no. 1 (1941): p. 52.
34. For a study of Theravada Buddhism during the colonial era, see Anne Ruth Hansen, *How to Behave: Buddhism and Modernity in Colonial Cambodia, 1860–1930* (Honolulu: University of Hawaii Press, 2007).
35. Susan Bayly, "French Anthropology and the Durkheimians in Colonial Indochina," *Modern Asian Studies* 34, no. 3 (2000).
36. See Pierre Bitard, "Le monde du sorcier au Cambodge," in *Le monde du sorcier* (Paris: Editions du Seuil, 1966), Marie Alexandrine Martine, "Elements de médicine traditionnelle khmer," *Seksa Khmer* 1, no. 6 (1983), Evelyn Porée-Maspero, "Notes sur les particularités du culte chez les cambodgiens," *BEFEO* XLIV, no. 2 (1950), Porée-Maspero, *Cérémonies privées.*
37. Byron J. Good, *Medicine, Rationality, and Experience: An Anthropological Perspective,* ed. Anthony T. Carter, Lewis Henry Morgan Lectures: 1990 (Cambridge: Cambridge University Press, 1994), p. 177.
38. Annick Guénel, "La lutte antivariolique en Extrême-Orient: Ruptures et continuité," in *L'aventure de la vaccination,* ed. Anne Marie Moulin (Paris: Fayard, 1996).
39. Ang Choulean, "Apports indiens à la médecine traditionnelle khmère," *Journal of the European Ayurvedic Society* 2 (1992), p. 110.
40. Bitard, "Le monde du sorcier au Cambodge"; Evelyn Porée-Maspero, "La cérémonie de l'appel des esprits vitaux chez les cambodgiens," *BEFEO* XLV, no. 1 (1951).
41. For a detailed analysis in the Khmer context, see François Bizot, *Le figuier à cinq branches: Recherche sur le bouddhisme khmer* (Paris: EFEO, 1976). See also, for Thailand, Phou Ngeun Souk-Aloun, *La médecine bouddhique traditionnelle en pays théravâda* (Limoges: Editions Roger Jollois, 1995). See also André Bareau, "Une representation du monde selon le tradition bouddhique," *Seksa Khmer* 1, no. 5 (1982).
42. This is also supported by Ebihara's observation that villagers did not distinguish the different religious traditions in their various ceremonies. May Ebihara, "Svay, a Rice Growing Village: 1958–1960" (PhD thesis, Columbia University, 1971), p. 363.
43. These examples are from copies of palm-leaf manuscripts provided to me by Olivier de Bernon at the Phnom Penh EFEO.
44. Among the ethnic Samre in western Cambodia, an entrant into the forest who did not offer the malevolent forest spirit Meas an offering of food risked stomach aches, joint pains, and fevers. Baradat, "Le Samrê ou Péâr: Population primitive de l'ouest du Cambodge," p. 53. Dr. Bernard Menaut also offers several examples of offending acts against different types of spirits in his article on Cambodian medicine in the 1920s. Bernard Menaut, "Matière médicale cambodgienne," in *Exposition coloniale internationale,* ed. Inspection Générale des Services Sanitaires et Médicaux de l'Indochine. (Hanoi: Imprimerie d'Extrême-Orient, 1931). One of the indigenous doctors of the Assistance Médicale provides a good description of a variety of genies, ghosts, and spirits that cause sickness, as well as a list of food taboos in his 1927 annual report for the district of Pursat. NAC RSC 116. One of the most comprehensive compilations of Khmer supernatural entities (and their role in health) in the more recent period is found in Ang Choulean's study, *Les êtres surnaturels.*
45. See the case of Neang Tuch, chapter 4. Individuals were known to be punished in the Khmer courts for "killing by witchcraft." For example, see NAC RSC 25602, where a man named Ok was found guilty of killing his neighbor's wife with a magical spell. The belief in magical manipulation of an individual's health or emotional well-being is still common today in many Southeast Asian countries, even among urban dwell-

ers. C. W. and Roy Ellen Watson, ed., *Understanding Witchcraft and Sorcery in Southeast Asia* (Honolulu: University of Hawaii Press, 1993).

46. Taboos, particularly food taboos, do not solely belong in the realm of Khmer medicine. A French doctor blamed the spread of the cholera epidemic in 1895 to the Cambodian habit of eating green mangos. NAC RSC 2175. Dr. Menaut, mentioned above, also believed that green mangos were implicated in the spread of conjunctivitis. NAC RSC 1248.
47. Wind, one of the four elemental *cato phut*, is a vital element of both Ayurvedic and Buddhist medicine. However, it seems to hold a particularly important place in Khmer medicine.
48. See Martine, "Elements de médicine traditionnelle khmer."
49. Moxibustion, another common form of home treatment, works on a similar premise.
50. Baradat, "Le Samrê ou Péâr: Population primitive de l'ouest du Cambodge," p. 60, Martin Piat, "Médecine populaire au Cambodge," *Bulletin de la Sociéte des Etudes Indochinoises* XL, no. 4 (1965).
51. See Bitard, "Le Monde du sorcier au Cambodge."
52. CAOM BIB 20037.
53. Ibid. See also André Souyris-Rolland, "Les procédés magiques d'immunisation chez les cambodgiens," *Bulletin de la Sociéte des Etudes Indochinoises* XXVI, no. 2 (1951).
54. See Bitard, "Le monde du sorcier au Cambodge.", F. G. Faraut, *Astronomie cambodgienne* (Phnom Penh, Saigon: Imprimerie F. H. Schneider, 1910).
55. Ok Chan, "Contribution à l'étude de la therapeutique traditionnelle au pays khmer: Thèse pour le doctorat en médecine" (MD thesis, Faculte de Médecine de Paris, 1955); Menaut, "Matière médicale cambodgienne"; Ritharasi Norodom, "L'évolution de la médecine au Cambodge" (MD thesis, Librarie Louis Arnett, 1929).
56. Aymonier, in his cataloguing of Khmer terms and beliefs, identifies healing properties ascribed to various common foods and objects (crab grease, tamarind paste, and frangipane, to name a few). Etienne Aymonier, *Notes sur les coutumes et croyances superstitieuses des cambodgiens* (Paris: CEDORECK, 1984 (1883)). Norodom and Chan also list medically significant plant and animal products in their medical theses. Chan, "Therapeutique traditionnelle"; Norodom, "L'évolution de la médecine." A later thesis by the pharmacist Phana Douk provides an extensive list of herbal products. Phana Douk, "Contribution à l'étude des plantes médicinales du Cambodge" (PhD thesis, Faculté de Médecine de Paris, 1965). Currently, the Khmer Association for Traditional Medicine continues these efforts, having published several catalogues in the post-UNTAC period. The nongovernmental organization Oxfam Novib produced a 1996 survey of medicinal plants used in the northeastern province of Ratanakiri. Other NGOs, particularly medical NGOs, are also collecting information.
57. NAC RSC 23452 and 28874. CAOM INDO RSC 235. The wide popularity of magically imbued healing water in the Mekong Delta region is mentioned in the writings of the seventeenth-century Jesuit missionary Alexandre de Rhodes. See Sjaak van der Geest, Susan Reynolds Whyte, and Anita Hardon, "The Anthropology of Pharmaceuticals: A Biographical Approach," *Annual Review of Anthropology* 25 (1996), p. 154.
58. Souyris-Rolland describes these tattoos in detail. Martel notes that only men have medico-magical tattoos, although she does not analyze the observation. Souyris-Rolland, "Les Procédés Magiques," Martel, *Lovea*, p. 56. Her observation is certainly borne out in contemporary observations of the population. Although tattoos are not common in the younger generation, faded tattoos can still be observed among the older generation of male Khmers.

59. Or of the French terms *docteur* and *médecin,* although the subtle difference in meaning between those two terms also complicates the issue.
60. Piat, "Médecine populaire au Cambodge."
61. Martine, "Elements de médicine traditionnelle khmer."
62. Norodom, "L'évolution de la médecine," conclusions. These differences could, of course, be due to radical changes in Khmer society in the fifty years between the two studies, as palm-leaf manuscripts are rapidly disappearing in the countryside.
63. Most notably, researchers at the Phnom Penh branch of the EFEO have been collecting and recording these documents from *wats* in the Cambodian countryside.
64. However, the Cambodian Association for Traditional Medicine has compiled several volumes of herbal medicinal recipes from these existing manuscripts and member records.
65. For example, in "L'évolution de la médecine," Norodom uses the transliterated word *mâ* (p. 47–48) as the translation for *médecin* in Khmer spiritual healing tradition. Written in the 1920s, his thesis uses inconsistent transliteration, and it is difficult to discern the original Khmer word. *Mâ* may also refer to the Thai word for doctor, although the reason for the slippage between Thai and Khmer here is unclear.
66. Rather, early colonial documents refer to the Western doctor using the generic honorific title for a man, *lok.*
67. For example, in June 1939 a laborer had a dream directing him to oversee the exhumation of the remains of two individuals from a local *wat* and their reburial under a specific banyan tree in Kampot. The *achaar,* being the brother-in-law of the laborer's patron, complied with the magical dream. By the end of the year, several Buddha images had been placed around the tree, and it had become the site of regular pilgrimages of the sick. NAC RSC 28944.
68. Scholars know of the existence of these hospitals through archaeological evidence. Inscribed steles describe these 102 hospitals built under Jayavarman VII's rule, and the ruins of several of these hospitals have been found. See Rethy K. Chhem, "Medicine and Culture in Ancient Cambodia," *The Singapore Biochemist* 13 (2000); George Coedès, "L'assistance medical au Cambodge à la fin du Xiie siecle," *Revue médicale française d'Extrême Orient* 19, no. 1 (1941), Claude Jacques, "Les édits des hopitaux de Jayavarman VII," *Etudes Cambodgiennes,* no. 13 (1968); Menaut, "Matière médicale cambodgienne." Jayavarman VII's hospitals are also discussed in the introduction to Choulean's and Hou's studies. Choulean, "Apports indiens"; Ngo Hou, "Les débuts de l'assistance médicale au Cambodge de 1863 à 1908" (MD thesis, University of Hanoi, 1953). Jayavarman VII is often considered the last great Angkorean king. He erected many of the monuments of the Angkor Wat complex, as well as rest houses along the raised road systems and temples in other parts of the kingdom. He ruled from 1181 to approximately 1218.
69. Menaut, "Matière médicale cambodgienne." It was based on the stele inscriptions at Say Fong. See also George Coedès, *The Indianized States of Southeast Asia* (Honolulu: University of Hawaii Press, 1971), pp. 170–76.
70. Don Bates, ed., *Knowledge and the Scholarly Medical Traditions* (Cambridge: Cambridge University Press, 1995). Bates's introduction to this edited volume provided a useful way of thinking about the differences between Khmer and European medicine outside of anthropological analytical models. He points out that Gnostic traditions (based on the knower) of medical learning lasted far longer in the East than in the West, where epistemic (based on the known) traditions came to dominate. Wall's

chapter on medieval texts categorizes three species of medical texts: scripture/sacred (ayurveda); classics/authorities (Chinese TM); and plain texts (Western medicine). Khmer palm-leaf manuscripts, collected and compiled yet not open to editing or assessment, would seem to fall in the middle category of such a classification.

71. Norodom, "L'évolution de la médecine," p. 46.
72. Jacques Léonard, *La médecin entre les pouvoirs et les savoirs* (Paris: Aubier Montaigne, 1981).
73. See the scholarship of Erwin Ackerknecht, George Weisz, Ann La Berge, Catherine Kudlick, Jacques Léonard, and Matthew Ramsey, among many others.
74. See chapter 3.
75. NAC RSC 9614.
76. NAC RSC 11197.
77. François Ponchaud, *La cathédrale de la rizière: 450 ans d'histoire de l'église au Cambodge* (Paris: Librarie Arthème Fayard, 1990). For more information on Catholic medicine, see chapter 5.
78. Lapeysonnie, *La médecine coloniale: Mythes et réalités* (Paris: Seghers, 1988), pp. 69–91. Le Corps de Santé de la Marine remained a separate entity.
79. Pierre Pluchon, *Histoire des médecins et pharmaciens de marine et des colonies* (Toulouse: Editions Privat, 1985), p. 186.
80. Philip D. Curtin, *Death by Migration: Europe's Encounter with the Tropical World in the Nineteenth Century* (Cambridge: Cambridge University Press, 1995).
81. This is, of course, a greatly oversimplified statement for a long and complicated process. The history of Western medicine and public health in Southeast Asia is still relatively sparse. For Indonesia, see Peter Boomgaard, "Dutch Medicine in Asia, 1600–1900," in *Warm Climates and Western Medicine: The Emergence of Tropical Medicine, 1500–1900*, ed. David Arnold (Atlanta: Rodopi, 1996), A. de G. M. van Heteren, M. J. D. Poulissen, A. de Knecht-van Eekelen, and A. M. Luyendijk-Elshout, eds., *Dutch Medicine in the Malay Archipelago 1816–1942* (Amsterdam: Rodopi, 1989). For Malaya, see Lenore Manderson, *Sickness and the State: Health and Illness in Colonial Malaya, 1870–1940* (Hong Kong: Cambridge University Press, 1996). For the Philippines, see Rodney Sullivan, "Cholera and Colonialism in the Philippines, 1899–1903," in *Disease, Medicine, and Empire: Perspectives on Western Medicine and the Experience of European Expansion*, ed. Roy MacLeod and Milton Lewis (New York: Routledge, 1988).
82. AIP Fonds Yersin. Yersin's 1892 travel journal.
83. Pluchon, *Histoire des médecins et pharmaciens*, p. 261.
84. Lucien Gaide and Henry Bodet, *L'Assistance Médicale et la protection de la santé publique*, ed. Indochine Française. Section des Services d'Intérêt Social. Inspection Générale des Services Sanitaires et Médicaux de l'Indochine, Exposition Coloniale Internationale (Hanoi: Imprimerie d'Extrême Orient, 1931).
85. Clavel, *L'Assistance Médicale Indigène en Indo-Chine: Organisation et fonctionnement* (Paris: Augustin Challamel, 1908). The decree is reproduced in full on pp. 7–12 of this report.
86. The title *médecin inspecteur* was shortly after changed to *inspecteur general*. For the sake of simplicity, I will simply use the term inspector general.
87. Clavel, *L'Assistance Médicale Indigène*, p. 3.
88. The health director for Cochinchina, the one region under direct rule, actually had the slightly different title of *directeur du service de santé* (*de la Cochinchine*).

89. I refer to the local health director of Cambodia simply as the health director. All other regional health directors will be referred to as the local health director of "Region A."
90. When the DLS first requested these tallies, many districts ordained on paper still had no medical facilities. Some had no doctors. See chapter 3 for more information.
91. The details of these changes will be discussed later in this study.
92. Clavel, *L'Assistance médicale indigène*, p. 4.

CHAPTER TWO

1. Andrew Cunningham and Perry Williams, "De-Centring the "Big Picture: The Origins of Modern Science and the Modern Origins of Science," in *The Scientific Revolution: The Essential Readings*, ed. Marcus Hellyer (Malden, MA: Blackwell, 2003); Roy MacLeod, "On Visiting the 'Moving Metropolis': Reflections on the Architecture of Imperial Science," in *Scientific Aspects of European Expansion*, ed. William K. Storey (Brookfield, VT: Variorum, 1996).
2. See Henri Brunschwig, *French Colonialism 1871–1914: Myths and Realities* (New York: Praeger, 1964).
3. The École Coloniale was formerly the École Cambodgienne, a school started informally by August Pavie in 1885. Pavie brought several elite Khmers to Paris to educate them on French mores, history, and language. The following year, the school's name was changed to the École Coloniale, and it started taking both Khmer and Vietnamese students. Soon it was used to educate young potential French colonial administrators as well as Indochinese natives. By 1889, indigenous students were separated out into the "native section." In 1914, the native section was discontinued and the school served as a preparatory school for future French colonial administrators. For a more detailed history of the institution, see William B. Cohen, *Rulers of Empire: The French Colonial Service in Africa* (Stanford: Hoover Institution Press, 1971).
4. The British Medical Association was a powerful professional lobby that worked for the professional interests of its members in both the UK and its colonies. K. David Patterson, *Health in Colonial Ghana: Disease, Medicine, and Socio-Economic Change, 1900–1955* (Waltham, MA: Crossroads Press, 1981).
5. AIP. Fonds Indochine. Letter from Bombay to Roux in Paris, April 23, 1897.
6. NAC RSC 12632. This doctor, Pannetier, also published a study on Khmer society.
7. The field of tropical medicine in France was also tied to military medicine. The main center for tropical medicine studies in France was the Military School of Tropical Medicine (also known as Le Pharo) in Marseilles.
8. Koch and Kitasato had previously worked together on tuberculosis. Koch was in India during the 1897–98 plague epidemic, and Kitasato worked in Hong Kong during the 1894 plague epidemic, but there is no evidence of collaboration.
9. See the classic works of Betts and Cady. Raymond F. Betts, *Assimilation and Association in French Colonial Theory 1890–1914* (New York and London: Columbia University Press, 1961); John F. Cady, *The Roots of French Imperialism in Eastern Asia* (Ithaca, NY: Cornell University Press, 1954). See also Andrew, "The French Colonial Party"; Brunschwig, *French Colonialism*; and Cohen, *Rulers of Empire*.
10. The development of this vaccine is also controversial, but most of the story occurs in Europe.
11. The use of antitoxin to induce limited immunity for diphtheria had come into use at the turn of the century, and had proven quite successful. Antitoxin induction of immunity (termed "passive immunity") was a substantially different means of "vac-

cination" from the use of attenuated disease virus ("active immunity"). For more details on the story of the development of the diphtheria antitoxin, see Evelynn Maxine Hammonds, *Childhood's Deadly Scourge: The Campaign to Control Diphtheria in New York City 1880–1930* (Baltimore: Johns Hopkins University Press, 1999).

12. In modern medical parlance, the term vaccine is sometimes used exclusively for an antigenic injection of attenuated virus or bacteria designed to induce active immunity. Antitoxin sera are distinct. This chapter uses the terms vaccine and vaccination as the colonial medical officials did: to denote both passive and active immunization by the microorganism or its byproducts.
13. "Clinical trial" in this period was simply experimental usage on (often uninformed) human subjects; the modern "clinical trial," with its randomized double-blind experiments, control groups, informed consent, and statistic analysis, was created after World War II. See Harry M. Marks, *The Progress of Experiment: Science and Therapeutic Reform in the United States, 1900–1990* (Cambridge: Cambridge University Press, 1997).
14. This situation was similar to the British situation in Burma during this period. See Atsuko Naono, "The State of Vaccination: British Doctors, Indigenous Cooperation, and the Fight against Smallpox in Colonial Burma" (PhD thesis, University of Michigan, 2005).
15. See two edited volumes on colonial medicine: David Arnold, ed., *Warm Climates and Western Medicine: The Emergence of Tropical Medicine, 1500–1900* (Atlanta: Rodopi, 1996); Roy MacLeod and Milton Lewis, eds., *Disease, Medicine, and Empire: Perspectives on Western Medicine and the Experience of European Expansion* (New York: Routledge, 1988).
16. The effectiveness of a vaccine hinges on the reliability and "fixed" nature of its attenuation. The smallpox virus was sometimes not sufficiently weakened during vaccine production. Rather than transmitting a milder form of the disease manifested by slight reactions and long-term immunity, such a vaccine induces the disease in its full virulence. One could argue that vaccination arm-to-arm with a possibly virulent form of smallpox is essentially a form of variolation.
17. Albert Calmette, "Compte rendu relatif à l'organisation et au fonctionnement de l'institut de vaccine animale crée à Saigon en 1891," (Saigon: Imprimerie Coloniale, 1891).
18. J. Nogue, "Missions de vaccine au Cambodge," *Annales d'hygiène et de médecine coloniales* (1898).
19. The undersecretary of colonies was the precursor to the French minister of colonies, created in 1894. Etienne was also to found the Colonial Party. See Cohen, *Rulers of Empire*; C. M. Andrew and A. S. Kanya-Forstner, "The French Colonial Party: Its Composition, Aims and Influence," *Historical Journal* 14, no. 1 (1971).
20. The Pasteur Institute had been founded in November 1888.
21. AIP. Fonds Indochine.
22. For information on Louis Pasteur and his efforts to turn theoretical scientific research towards economic and socially useful ends, see, for example, René Dubos, *Louis Pasteur, Free Lance of Science*, 2 ed. (New York: Da Capo Press, 1960); Gerald L. Geison, *The Private Science of Louis Pasteur* (Princeton, NJ: Princeton University Press, 1995); Bruno Latour, *The Pasteurization of France*, trans. Alan Sheridan and John Law (Cambridge, MA: Harvard University Press, 1988).
23. AIP. Fonds Armand Corre.

24. For its first two years of operation, Calmette's institution did not have an official name. In its first year of operation, Calmette referred to it as the Animal Vaccine Institute. The inspector general of health services referred to it as the Saigon Bacteriology Institute.
25. See also Annick Guénel, "The Creation of the First Overseas Pasteur Institute, or the Beginning of Albert Calmette's Pastorian Career," *Medical History* 43 (1999). Guénel provides more of Calmette's background before his arrival in Indochina, arguing Calmette's active role in setting the agenda of the Saigon Pasteur Institute and all bacteriological research in Indochina. She also provides a detailed description of several of his research programs during his short stint in Saigon.
26. See Jack Moseley, "Travels of Alexandre Yersin: Letters of a Pastorian in Indochina, 1890–1894," *Perspectives in Biology and Medicine* (1981). Moseley provides some interesting excerpts of Yersin's letters to his mother during his initial travels in Indochina. Other correspondence in the Pasteur Institute files contradicts some of Yersin's news to his mother.
27. This colonial health service is the Assistance Médicale, which would be decreed in 1905 for Indochina.
28. AIP Fonds Calmette and Fonds Simond. Personal letters.
29. AIP. Fonds Indochine.
30. The competition between military and government-appointed doctors in the French colonies was not unique to Indochina. For instance, the royally appointed doctors of the colony of Saint-Domingue (modern-day Haiti) were in direct competition with the doctors of the French navy in the eighteenth century. James McClellan, "Science, Medicine, and French Colonialism in Old-Regime Haiti," in *Science, Medicine, and Cultural Imperialism* (New York: St. Martin's Press, 1991).
31. AIP. Fonds Indochine. Letter from Calmette in Paris to Noël Bernard in Saigon, January 18, 1924.
32. Ibid. Correspondence between Bernard and Calmette, 1924.
33. NAC RSC 12632.
34. Le Pharo is the common name for the Military School of Tropical Medicine in Marseilles.
35. AIP. Fonds Simond. Letter from Simond to Calmette, June 30, 1910.
36. It appears that Bernard only made one brief research trip to Cambodia, in 1923 to study the "plague of newborns."
37. AIP. Fonds Indochine.
38. By 1937, the Health Corps was embroiled in another debate with the Pasteurians. Having "lost" the Hué lab, the corps had been trying for several years to obtain GGI approval to create an independent cholera vaccine facility at Hué, again arguing that it could produce a greater quantity of vaccine at a lower price for its own use. The Pasteur Institute vigorously fought these efforts, claiming once again that the nonspecialists would be unable to correctly build and monitor such a facility. The GGI sided with the Pasteurians, although whether for financial or technical reasons is unclear. A second Hué lab was never built. AIP Fonds Indochine.
39. This would be the beginning of the third global bubonic plague pandemic in history. When it finally ended in the early twentieth century, it had spread to every populated continent on the globe. Myron Echenberg, "Pestis Redux: The Initial Years of the Third Bubonic Plague Pandemic, 1894–1901," *Journal of World History* 13, no. 2 (2002).

40. Yersin was so out of favor with the English doctor Lowson that he was unable to legally obtain the corpses of plague victims for research purposes. He resorted to bribing longshoremen who had been hired to bury these corpses to let him take samples before the dead were buried. AIP. Fonds Yersin. Report to the GGI, September 1894.
41. Initially, Kitasato's claims to discovery of the plague bacillus increased his political power in Japan. For more on Kitasato's research career, see James R. Bartholomew, *The Formation of Science in Japan: Building a Research Tradition* (New Haven: Yale University Press, 1989).
42. AIP. Fonds Yersin. Today the Latin name of the disease is *Yersinia pestis,* in honor of Yersin.
43. AIP. Fonds Indochine.
44. Letter from French Consulate in Hong Kong to GGI, November 12, 1896. AIP. Fonds Indochine.
45. The two sera differed in preparation. The Yersin serum was made of heat-killed virulent cultures of the bacillus combined with its toxin (similar in theory to the passive immunity induced by the diphtheria antitoxin). The Haffkine serum contained the heat-killed bacillus combined with phenic acid. Haffkine's treatment had no toxin, and it also required injection of much smaller quantities, but it produced a greater negative reaction.
46. Yersin and Haffkine were also competing with a Dr. Alessandro Lustig. Affiliated with a research institute in Florence, Lustig also performed several thousand experimental plague vaccinations during this epidemic.
47. Haffkine had performed large-scale experimental vaccinations in India before. In 1893 he had performed more than thirty thousand cholera vaccinations in Calcutta. For more information on Haffkine's cholera experiments, see Ilana Löwy, "From Guinea Pigs to Man: The Development of Haffkine's Anticholera Vaccine," *Journal of the History of Medicine and Allied Sciences* 47, no. 3 (1992).
48. Simond himself had little faith in the value of the serum. He also strongly disliked Yersin, and encouraged Roux to rename the serum, should it be perfected.
49. AIP. Fonds Indochine. Letters from Yersin to Roux.
50. Ibid. Letters from Yersin to Roux.
51. AIP. Fonds Indochine. Letter from Yersin to Roux, July 30, 1898.
52. Ibid.
53. He went so far as to theorize that ants, which ate the food waste beneath the cages of the laboratory monkeys, must have carried the disease outside the lab. Yersin was not aware of the experiments of his compatriot Simond on fleas and the plague until October of that year.
54. A hill station is a sort of health spa for colonial administrators and soldiers. Scientists believed that periodic rests at these stations, located in regions that more closely resembled the European climate, allowed functionaries to recover from the debilitating effects of tropical climates. Because they were usually located in the mountains of the colony itself, hill stations also enabled the colonials to periodically escape the tropical climates without the expense of a round-trip ticket to the metropole. Cambodia's only hill station was Bokor (near Kampot), created in the 1920s.
55. AIP. Fonds Indochine. Letter from Yersin to Roux, October 5, 1898.
56. Ibid. Letter from Yersin to Roux, December 5, 1898.
57. AIP. Fonds Yersin. *Le petit tonkinois,* June 16, 1903.

58. AIP. Fonds Indochine.
59. See, for example, Henri H. Mollaret and Jacqueline Brossollet, *Alexandre Yersin: 1863–1943, Un Pasteurien en Indochine* (Paris: Belin, 1993). This work contends that the links between the plague epidemic and the Nha Trang laboratory were constructed for political reasons, rather than what would seem to be the reality: that these links were disavowed for political reasons.
60. It is possible that the fate of Haffkine served as a cautionary tale for Yersin. Certainly, from Yersin's perspective, the two men were in competition. Haffkine's downfall may have encouraged Yersin's withdrawal from his own vaccine work into the much less controversial field of agronomy.
61. As mentioned at the beginning of this chapter, the ownership of the existing basic research laboratory that was originally to be connected with the medical school had been transferred to the Pasteur Institute. Thus, this institute was created in 1904 as a research institute linked to the new École de Médecine. It, too, would become a full-fledged Pasteur Institute in 1926.
62. This was not the first time colonial prisoners had been experimental subjects for the cholera vaccine; Haffkine had experimented with his vaccine during an 1891 outbreak at the Gya jail in India. E. H. Hankin, *Cholera in Indian Cantonments and How to Deal with It* (Allahabad: Pioneer Press, 1895), p. 31.
63. AIP. Fonds Simond. Report of Gauducheau to Simond, June 6, 1916.
64. AIP. Fonds Simond. Cablegram from Saint-Chaffrey to GG in Paris, May 17, 1916.
65. AIP. Fonds Simond. "Cholera Journals," May and June 1916.
66. AIP. Fonds Simond. Report by Gauducheau to minister of colonies, December 16, 1916.
67. In the second decade of the twentieth century, epidemiologists were regularly using relatively sophisticated statistical techniques such as multifactorial analysis, stratification, and adjustment for potential confounds.
68. Tetanus, sometimes called lockjaw, is caused by an exotoxin secreted by the *Clostridium tetani* spore. Kitasato is credited with isolating the spore in 1889. The disease causes muscle spasms and convulsions and is often fatal if untreated. It is not contagious between individuals. An antitoxin for the disease would not be developed until the mid-1920s. By the 1930s, a combined vaccine for tetanus and typhoid was in use in Indochina (see beginning of this chapter).
69. This incident is highly suggestive of the public relations tactics of Louis Pasteur himself. See Geison, *The Private Science of Louis Pasteur*.
70. AIP. Fonds Indochine. Report by Bernard to Comte, June 16, 1922.
71. Ibid.
72. The use of prisoners, military conscripts, and other co-opted groups to test new treatments is a much more problematic issue directly tied to oppression and exploitation of individuals without choice. The guidelines for informed consent and acceptable human experimentation would not become central to the language of responsible scientific conduct until after the shocking Nazi excesses of World War II. See Paul Weindling, "The Origins of Informed Consent: The International Scientific Commission on Medical War Crimes, and the Nuremburg Code," *Bulletin of the History of Medicine* 75 (2001). These vaccine researches in India and Indochina were undoubtedly exploitation. But it is difficult to contextualize the attitudes of these scientists towards this human exploitation, and I hesitate to characterize it specifically as an issue of racism, colonial power, state power, scientific arrogance, or some combination thereof without careful qualification in comparison with contemporary metropolitan trends.

73. NAC RSC 23033.
74. NAC RSC 23033. Letter from Guy Tromeur to the *La Presse Indochinoise*, March 1932.
75. Local villagers successfully petitioned for Dr. Fabry's removal as the medical chief of the district of Kompong Cham.
76. NAC RSC 23033.
77. See Erwin Heinz Ackerknecht, *Medicine at the Paris Hospital, 1794–1848* (Baltimore: Johns Hopkins Press, 1967); Andrew R. Aisenberg, *Contagion: Disease, Government, and The "Social Question" in Nineteenth-Century France* (Stanford, CA: Stanford University Press, 1999); Oliver Faure, *Les français et leur médecine aux XIXe siècle* (Paris: Belin, 1993); Martha L. Hildreth, "Doctors and Families in France, 1880–1930: The Cultural Reconstruction of Medicine," in *French Medical Culture in the Nineteenth Century*, ed. Ann La Berge and Mordechai Feingold (Atlanta: Editions Rodopi, 1994); Ann La Berge and Mordechai Feingold, eds., *French Medical Culture in the Nineteenth Century* (Atlanta: Editions Rodopi, 1994); Ann F. La Berge, *Mission and Method: The Early Nineteenth-Century French Public Health Movement* (Cambridge: Cambridge University Press, 1992); Jacques Léonard, *La médecin entre les pouvoirs et les savoirs* (Paris: Aubier Montaigne, 1981); Harry W. Paul, *From Knowledge to Power: The Rise of the Science Empire in France, 1860–1939* (Cambridge: Cambridge University Press, 1985). See also Weisz's article for a succinct definition of medical professionalization. George Weisz, "The Politics of Medical Professionalization in France 1845–1848," *Journal of Social History* 12, no. 1 (1978).
78. See the collection of articles in MacLeod and Lewis, eds., *Disease, Medicine, and Empire*. For India, see Mark Harrison, "The Tender Frame of Man: Disease, Climate, and Racial Difference in India and the West Indies, 1760–1860," *Bulletin of the History of Medicine* 70, no. 1 (1996); and *Public Health in British India: Anglo-Indian Preventive Medicine 1859–1914* (Cambridge: Cambridge University Press, 1994).
79. NAC RSC 1355. Letter to RSC, March 23, 1926.
80. Ibid. Letter to RSC, April 2, 1926.
81. NAC RSC 12635. The Khmer Mixture will be discussed further in chapter 3.
82. NAC RSC 77. Annual report of the Hygiene Service, 1926.
83. NAC RSC 1729. Kampot annual report, 1928.
84. NAC RSC 2479. Annual report, 1928. Menaut DLS.
85. NAC RSC 127. Annual report, Takeo, 1930. A *khum* is a large village or a group of small villages (*phum*).
86. NAC RSC 136.
87. NAC RSC 31881.
88. NAC RSC 12632.
89. NAC RSC 1344. The Pasteur Institutes of Indochina experimented with modeling their fundraising efforts on the metropolitan laboratory. The Parisian Pasteur Institute was in large part funded by public donations.
90. NAC RSC 1066.
91. NAC RSC 7638.
92. NAC RSC 31881.
93. For Europe and America, see James Colgrove, "'Science in a Democracy': The Contested Status of Vaccination in the Progressive Era and the 1920s," *Isis* 96, no. 2 (2005); Nadja Durbach, "Class, Gender, and the Conscientious Objector to Vaccination, 1898–1907," *The Journal of British Studies* 41, no. 1 (2002); Naomi Williams, "The Implementation of Compulsory Health Legislation: Infant Smallpox Vaccination

in England and Wales, 1840–1890," *Journal of Historical Geography* 20 (1994). It is interesting to note that today there is a vocal and vehement anti-vaccination campaign in the United States and England, linked to parental fears of autism.

94. Académie de Médecine, "Rapport générale présenté à M. le ministre de l'intérieur par L'Académie de Médecine sur les vaccinations et revaccinations pratiquées en France et dans les colonies pendant l'année 1899," (Melun: Académie de Médecine. Imprimerie Administrative, 1899).
95. NAC RSC 31840.
96. See Penny Edwards, *Cambodge: The Cultivation of a Nation, 1860–1945* (Honolulu: University of Hawaii Press, 2006).
97. See John Sydenham Furnivall, *Colonial Policy and Practice: A Comparative Study of Burma and Netherlands India* (New York: New York University Press, 1956).
98. NAC RSC 709.
99. NAC RSC 1315. Annual report of Devy for Phnom Penh municipality, 1910.
100. Ibid. The comparable advantages and disadvantages of the two vaccines were never entirely resolved throughout the French colonial period.

CHAPTER THREE

1. Jacques May, *Un médecin français en Extrême-Orient* (Paris: Editions de la Paix, 1951), pp. 147–48.
2. Henri Brunschwig, *French Colonialism 1871–1914: Myths and Realities* (New York: Praeger, 1964), p. 167.
3. NAC RSC 1322.
4. NAC RSC 12505. The Pnongs are an ethnic mountain tribe, often portrayed as quite primitive. Among urban Khmer, "Pnong" is an insult equivalent to calling someone a barbarian or savage.
5. Francis Garnier, *Travels in Cambodia and Part of Laos: The Mekong Exploration Commission Report (1866–1868)*, vol. 1, trans. Walter E. J. Tips (Bangkok: White Lotus Press, 1996 [1885]).
6. NAC RSC 9614.
7. E. Maurel, "Mémoire sur l'anthropologie des divers peuples vivant actuellement au Cambodge," *Mémoires de la Société d'Anthropologie de Paris*, series 2, no. T.4 (1893). Disturbingly, Cambodia's Resident Supérieur Fourès gave Maurel permission to excavate the skulls of recently beheaded Khmers for his 1885 craniometry study. The doctor received permission to exhume these heads from a common grave six months after the men's executions. Maurel would note the "dedication" of his fellow military doctors, who were forced to stop and vomit several times during the exhumation due to the condition and odor of the graves.
8. The term "ambulance" usually denotes a mobile or makeshift military hospital. An ambulance in the colonial context was often simply any building where the doctor regularly saw patients.
9. During the first severe plague epidemic in 1907, the colonial government instituted some draconian measures. The Chinese congregation lodged protests about the extended closing of buildings and the transfer of the sick to the lazarets, as well as about troublesome disinfection measures such as fumigation by sulfurous gas and whitewashing. They also objected to hospitalization of their peers in the Mixed Hospital. Using their commercial clout, the "Celestes," as they were whimsically referred to by the administration, obtained a major concession from the medical administration when it allowed them to open a Chinese Hospital in 1907. The AM enforced

health regulations within the service, performed regular inspections, and maintained ultimate control of the institution. In exchange for these terms, the Chinese were permitted under their own funding and organization to build, staff, and run the hospital. See Sokhieng Au, "Indigenous Politics, Public Health and the Cambodian Colonial State," *South East Asia Research* 14, no. 1 (2006).

10. In 1911, budget cuts and personnel shortages forced the AM to contract the twelve outlying districts to ten. By 1915, with medical staff being funneled to the European war front, several ambulances closed down and only seven administrative medical districts remained. After World War I, the number of provincial medical districts slowly returned to prewar levels. Due to the increased importance of Angkor Wat as a tourist destination, Siem Reap was separated from Battambang as its own medical district in 1925.
11. Post le Rolland was created shortly after the rebellion of the mountain-dwelling ethnic Pnongs in northeastern Cambodia. The rebellion, which was viciously repressed by French soldiers, and the sensationalized newspaper coverage of the lack of medical care for sick soldiers and civilians impelled the administration to improve the health care situation in this remote post of Haut Chhlong. See *La Presse Indochinoise* for 1934 and 1935; *Echo du Cambodge*, March 1935 and February 1936; NAC RSC 1200, and 15238. Although it was technically declassified as a medical district after two years, separate medical statistics for the post would be kept for the rest of the decade.
12. NAC RSC 26681.
13. NAC RSC 1323.
14. AIP Fonds Simond.
15. In 1902, the first candidate from Cambodia admitted to the École was a young man by the name of Ferdinand Amphon. Likely métis, Ferdinand never completed the program. NAC RSC 2024.The first indigenous médecin posted in Cambodia was Sonn, of Khmer origin. He almost did not complete his studies, having been kicked out of the Ecole in 1909 for "fraternizing with an Annamite girl." He was put on probation and finished his training at the Mixed Hospital in Phnom Penh before taking and passing his exams to become an médecin auxilaire. His "Annamite girlfriend" Nguyen thi Lieng, was an ethnic Vietnamese born in Cambodia. Her career as an AM midwife will be briefly touched upon in the Chapter Five. NAC RSC 12617, 19034, 25675. For more on the Ecole de Médecine, see Chapter Two.
16. Indigenous *médecins* who billed themselves as "*docteurs*" were penalized by the administration. This was not, however, strictly an issue of distinguishing French from native. Indochinese men who finished a degree in the Faculté de Médecine of Paris were permitted the title of *docteur*. Indochinese men who were naturalized French, yet did not hold a degree from an accredited Université d'Etat, could not use the title. NAC RSC 605, 5925, 31895, 31902. In 1931, the École was reorganized with the first overseas Faculté de Médecine, permitting students to obtain the degree of *docteur* in Hanoi. Lapeysonnie, *La médecine coloniale: Mythes et réalités* (Paris: Seghers, 1988), p. 246.
17. NAC RSC 3143, 7634.
18. Some had been admitted with one or two years of *collège* (middle school) education; many had only an *école primaire* (primary school) education. NAC RSC 429.
19. NAC RSC 1254.
20. NAC RSC 2117.
21. NAC RSC 11996.

22. NAC RSC 7216.
23. In 1934, in the developing tourist town of Siem Reap, a distraught Khmer man brought his pregnant wife in to see Ketmoung. She was in distressed labor, the birth stalled. Ketmoung entered into a haggling session with the man in the presence of the suffering woman, getting him to agree to an exorbitant fee before aiding her. "Aid" was not the most appropriate term, as he proceeded to rip the nearly full-term infant out of its mother in pieces, much to the horror of the father. The father then refused to pay the agreed-upon fee—a fee that was also technically against regulations for AM staff. Ketmoung continued to harass him for payment until the father took the story to the newspapers. Occurring in the tourist town of Siem Reap, the scandal forced the AM to quickly recall Ketmoung to Phnom Penh.
24. NAC RSC 7577, 33461.
25. NAC RSC 7577. Ketmoung would get into further trouble after he was reposted, accusing an *infirmier* of gross negligence and a French officer of putting a gun to his head. He was intensely unpopular among both the medical staff and his patients, and the French found his accusations to be without merit. Further, he was deemed medically negligent for the very act of which he had accused his *infirmier*. He was still not dismissed.
26. *Infirmier* can be translated into English as "nurse"; however, *infirmiers* in the colonial context were also orderlies, translators, assistants, and guards. They were predominantly male. In order to avoid confusion with the English connotation of nurse, I retain the French word.
27. Major restructurings of the *infirmier* cadre were in 1915–16 (NAC RSC 5471), 1922 (NAC RSC 1249, 1254, 14641), 1925 (NAC RSC 2238), and 1934 (NAC RSC 11996).
28. NAC RSC 14439, 1322. Ritharasi Norodom, "L'évolution de la médecine au Cambodge" (M.D. thesis, Librarie Louis Arnett, 1929).
29. NAC RSC 598 and 429.
30. NAC RSC 78.
31. See, for example, NAC RSC 1561. The health director talks about the honor attached to different types of jobs, and the lack of "honor" attached to many of the tasks that a Western doctor performed.
32. Pondichery was a French colonial enclave in India.
33. NAC RSC 21239, and annual reports.
34. NAC RSC 605, and annual reports.
35. NAC RSC 2119. Letter from Inspector General Hermant to health director, June 13, 1933.
36. This was an unconvincing argument, since the attempts to admit inadequately educated Khmer candidates in the first fifteen years of the École's history had proven unsuccessful. Most candidates simply did not or could not complete their medical education, in part because they did not have the French language skills to follow instruction.
37. NAC RSC 11996. Letter from RSC Silvestre to the governor general, February 6, 1933.
38. David Chandler, *A History of Cambodia*, 3rd ed. (Boulder: Westview Press, 1996), p. 160.
39. The health director under the Vichy colonial government submitted a detailed plan for a Buddhist Medical School in Phnom Penh, developed in cooperation with the newly formed Indigenous Consultative Assembly in 1941. The school remained in

the planning stages, due to the unstable political situation at the time. NAC RSC 9164.

40. NAC RSC 11996.
41. NAC RSC 2117.
42. See, for example, Sok Khin, *L'annexion du Cambodge par les vietnamiens au XIX siècle* (Paris: Editions You-Feng, 2002). This book is a translation of a series of long poems by the monk Batum Baramey Pich, relating the experience of the nineteenth-century Vietnamese interregnum. The translator claims that Pich's poetry kept the atrocities committed by the invading armies of Siam and Vietnam alive in local Khmer memory. Although the Thais are not portrayed positively, the Vietnamese atrocities recounted in these poems are nearly fantastic , including the decapitation of Khmer children to use as ballasting rocks for cauldron fires to boil other Khmer children alive.
43. NAC RSC 11533, 17815, 17836, 31807, 31880.
44. NAC RSC 1250.
45. Ibid.
46. The tangential information I reviewed on these work sites indicates considerable mortality among the conscripted workers, and very little available health care.
47. NAC RSC 12624.
48. NAC newspaper collection. *Echo du Cambodge*, April 18, 1925.
49. NAC RSC 12632. Vaccine report of Dr. Denier, December 5, 1911.
50. NAC RSC 98.
51. NAC RSC 1315. Kompong Thom annual report, 1910.
52. NAC RSC 108.
53. NAC RSC 605. The impetus for the study was the lack of French medical influence in Indochina on the local population in 1938. More than twenty years after the formation of the AM, the health director noted that still only an estimated 10 percent of the native population ever had recourse to French medicine, when the AM's purpose as originally conceived was to reach the entire native population. The health director noted that, too often, the medical staff blamed this low number on the lack of budget. However, he observed, it simply wasn't possible to increase the budget ten times over; thus they had to look to other causes and solutions.
54. The Holman snake is actually a French corruption of the Khmer term for the snake, the Hanuman snake (sometimes pronounced Holomon). Hanuman is the Hindu monkey demigod, a prominent figure in the *Ramayana*. The vernacular English term for this snake is pit viper, its scientific genus *Trimeresurus*.
55. NAC RSC 7808.
56. NAC RSC 630, Kompong Chhnang annual report, 1923.
57. NAC RSC 134.
58. NAC RSC 78.
59. Bonnigal would become municipal doctor in Phnom Penh a few years later, a position he would hold for several years.
60. NAC RSC 12632. Letter from Pujol to RSC, January 4, 1909.
61. AC RSC 12632.
62. NAC RSC 1315. Kompong Thom annual report, 1920.
63. NAC RSC 2414.
64. NAC RSC 78.
65. NAC RSC 7808.
66. NAC RSC 78. Annual ensemble report, 1927.

67. NAC RSC 605, 2107, 9291.
68. Nicholas Thomas, *Entangled Objects: Exchange, Material Culture, and Colonialism in the Pacific* (Cambridge, MA: Harvard University Press, 1991), p. 4.
69. Sjaak van der Geest, Susan Reynolds Whyte, and Anita Hardon, "The Anthropology of Pharmaceuticals: A Biographical Approach," *Annual Review of Anthropology* 25 (1996): pp. 168–70. See also Bryan Pfaffenberger, "Social Anthropology of Technology," *Annual Review of Anthropology* 21 (1992).
70. Pfaffenberger, "Social Anthropology of Technology," p. 501.
71. NAC RSC 1334, 11469. This was the case with the Chhlong ambulance built in early 1927 in the medical district of Kratie, and the Kompong Som dispensary in the Kampot district in 1930.
72. NAC RSC 1333, 1336.
73. NAC RSC 1333.
74. NAC RSC 1332. Letter from health director to resident of Kratie, January 17, 1928.
75. NAC RSC 1336.
76. NAC RSC 17834.
77. Ibid. Letter from health director to RSC, July 17, 1921.
78. NAC RSC 2118, AM annual report, 1939. The number of doctors would actually decrease slightly in 1940, with mobilization for the war. NAC RSC 2119.
79. NAC RSC 2117. AM annual report, 1938. The 1939 report does not provide estimated population.
80. NAC RSC 2118.
81. NAC RSC 29748. This file contains hundreds of these small sheets of information, collected over a period of a few months.
82. NAC RSC 32644.
83. NAC newspaper collections. *Echo du Cambodge*. See June 26, 1936 issue.
84. NAC newspaper collections. *Le Populaire*, September 12, 1935. For a description of the interrogation process of these patients, see *La Dépêche*, September 25, 1935.
85. NAC RSC 32551.
86. NAC RSC 2280.
87. NAC RSC 1315.
88. NAC RSC 25784. The instructions are all quite practical. For instance, for a drowning victim the brochure presents a very basic sort of CPR. "First the rescuer pulls the victim out of the water. If the victim is not breathing, the rescuer should remove the victim's clothes, put him on his back and hit him in the stomach and heart. Then, take a wet towel and smack the victim. Turn the victim's head to the right side and then blow air into his mouth in order to make him breathe. Then make a fire and create smoke so that the victim chokes out any water swallowed. Then turn him left and then right, and then turn his head and blow in his mouth again, and repeat until the person regains consciousness."
89. NAC RSC 12607.
90. Further, the Chastang potion, while developed by a local French pharmacist, seems to also have been based on an indigenous remedy.
91. Ibid.
92. NAC RSC 2485. Provides a record of the amount of Chastang Potion used for the 1908 cholera epidemic in Phnom Penh. See also chapter 2.
93. NAC RSC 12619. Compare to the stance of the governor of Tuk Khleang in Prey Veng, who in 1915 of his own volition also translated French instructions for the

treatment of cholera; he however distributed instructions on the Chastang Potion. NAC RSC 15254.

94. NAC newspaper collections. *Echo du Cambodge.*
95. These insurrections flared up immediately after King Norodom was coerced into signing an 1884 treaty transferring most of the power of government to the French. They lasted until 1886, when Norodom was persuaded or coerced again by French officials to urge the various insurrectionists to cease fighting. See chapter 1.
96. NAC RSC 9614. However, Monnais-Rousselot identifies the first infirmary in Cambodia as one built in 1863 by Doudart Lagree at Kompong Luong (near Oudong) and manned by Dr. Hennecart to serve the French and the Khmer royal family. See Laurence Monnais-Rousselot, *Médecine et colonisation: L'aventure indochinoise* (Paris: CNRS Editions, 1999), pg. 86. The medical infirmary was transitory, since Oudong was abandoned as the seat of government in 1866.
97. NAC RSC 9614. Maurel's journal.
98. NAC RSC 11196.
99. NAC RSC 9614.
100. NAC RSC 2486. AM annual report, 1908.
101. Charles E. Rosenberg, *The Care of Strangers: The Rise of America's Hospital System* (New York: Basic Books, Inc., 1987), p. 7.
102. Oliver Faure, *Les français et leur médecine aux XIXe siècle* (Paris: Belin, 1993), pp. 30–32, 88–94.
103. NAC RSC 2118.
104. NAC RSC 2379.
105. NAC RSC 2486.
106. NAC RSC 24015. Letter from Dr. Bonneau in Sisiphon to Battambang doctor, 1914.
107. NAC RSC 1315.
108. NAC RSC 4554.
109. NAC RSC 26686.
110. This is a commonly held Khmer belief even today.
111. NAC RSC 7637.
112. NAC RSC 7878.
113. For example, *La Dépêche* October 4, 1935; *La Presse Indochinoise* May 17, 1935; NAC RSC 11996.
114. NAC RSC 9116.
115. As is still common in Cambodia, most businesses closed for several hours during midday when the heat was strongest. At businesses that did not close, workers often napped in place.
116. NAC RSC 9116. Rigaud's temper and his marriage to a Khmer woman are fortuitous for the historian, for they provide a glimpse of the voices missing in the historical record. As was apparently the case with Oum's brother and husband, it is unlikely that the indigenous population in other instances would have dared voice their complaints with the medical authorities.
117. NAC RSC 9116. Personal letter from Rigaud to Silvestre, November 26, 1934.
118. "To go to Canossa" is a figure of speech denoting being forced to bow in humiliation to an opponent. It would seem that a higher authority, probably the GGI, intervened in the situation.
119. See chapter 2.

120. How the indigenous population felt about these changes in regulation is unclear since these newspapers were read only by the French-speaking population. *La Presse Indochinoise* and *L'Aurore* seemed to have a vendetta against each other in the early 1930s; and both at times attacked the running of the Mixed Hospital, and at other times criticized the other's attacks. See, for example, *L'Aurore* from September 25, 1935, or most issues of *La Presse Indochinoise* from 1934 and the first half of 1935. *La Dépêche* (December 1935) and *Le Populaire* (September 1935) both ran very positive articles about the appointment of Simon as the new health director.
121. NAC RSC, AM annual reports, 1931–1935. See also NAC newspaper clippings, *La Presse Indochinoise.*

CHAPTER FOUR

1. The quote is from NAC RSC 7637. These charts are compiled with data from district annual reports. Because often the data was changed, omitted, or condensed in the ensemble report for the country, I chose two representative medical districts with relatively complete records for the number of vaccinations as well as for the number of smallpox deaths.
2. NAC RSC 2115, 25607.
3. In 1925 the inspector general would note that the number of smallpox deaths was rising rather than falling in the country. NAC RSC 1350.
4. See Andrew R. Aisenberg, *Contagion: Disease, Government, and the "Social Question" in Nineteenth-Century France* (Stanford, CA: Stanford University Press, 1999); David S. Barnes, *The Making of a Social Disease: Tuberculosis in Nineteenth-Century France* (Berkeley: University of California Press, 1995); Ann F. La Berge, *Mission and Method: The Early Nineteenth-Century French Public Health Movement* (Cambridge: Cambridge University Press, 1992); Charles E. Rosenberg, *The Cholera Years* (Chicago: University of Chicago Press, 1962); Barbara Rosenkrantz, *Public Health and the State: Changing Views in Massachusetts, 1842–1936* (Cambridge, MA: Harvard University Press, 1972). Rosenkrantz notes of the public health movement in Massachusetts that we must be mindful that the public health expert is not "purposefully hiding social controls under the camouflage of interest in public welfare."
5. Anne Marie Moulin, "Bacteriological Research and Medical Practice," in *French Medical Culture in the Nineteenth Century*, ed. Ann La Berge and Mordechai Feingold (Atlanta: Editions Rodopi, 1994), p. 342.
6. Penny Edwards, *Cambodge: The Cultivation of a Nation, 1860–1945* (Honolulu: University of Hawaii Press, 2006), p. 17.
7. NAC RSC 7637. Phnom Penh Municipality annual report.
8. Erwin H. Ackerknecht, "Hygiene in France, 1815–1848," *Bulletin of the History of Medicine* 22, no. 2 (1948).
9. Baldwin defined this "geoepidemological" location both as the geographic position of the country in relation to a policy learning curve regarding new invading epidemics (countries further away from the original source of the disease had the leisure of watching and learning as their first-affected European brethren mismanaged control of the disease) and to the relative porousness of a state's borders to specific epidemic threats. He did, however, note that some traditions of public health were learned. Peter Baldwin, *Contagion and the State in Europe, 1830–1930* (Cambridge: Cambridge University Press, 1999).
10. Gwendolyn Wright, *The Politics of Design in French Colonial Urbanism* (Chicago: University of Chicago Press, 1991); Paul Rabinow, *French Modern* (Cambridge, MA: MIT

Press, 1989). These authors point to colonial administrators, and most specifically to Herbert Lyautey in Morocco, as using the colony as a sort of laboratory for modernity. French urban planners and architects, due to their wider range of powers in the colonial context, had a freer field to experiment with their own vision of the modern city than did their counterparts in France. Although part of these authors' formulation of the colonial laboratory deals with the sanitary engineers, and thus hygienic sciences, for the most part their analyses are focused on experimentation with *cultural* norms and forms.

11. See William Coleman, *Yellow Fever in the North: The Methods of Early Epidemiology* (Madison: University of Wisconsin Press, 1987); Baldwin, *Contagion and the State,* chapters 2, 3 and 4.
12. John Sydenham Furnivall, *Colonial Policy and Practice: A Comparative Study of Burma and Netherlands India* (New York: New York University Press, 1956).
13. Dipesh Chakrabarty, "Subaltern Studies and Postcolonial Historiography," *Nepantla: Views from the South* 1, no. 1 (2000), p. 21. This formulation of subaltern scholarship can be traced back to the theories of Gramsci and even earlier to the writings of Marx and Engels, theorists who believed that the powerless were subject to multiple sets of related but at times competing (hegemonic) ideologies originating from the state and the (indigenous) elite.
14. NAC RSC 1729.
15. Louis Chevalier, *Le chólera: La première épidémie du XIX siècle* (La Roche-Sur-Yon: Imprimerie Centrale de L'Ouest, 1958); Catherine Jean Kudlick, *Cholera in Post-Revolutionary Paris: A Cultural History* (Berkeley: University of California Press, 1996). Both of these authors are primarily concerned with how the bourgeoisie defined its perception of cholera in terms of distancing itself from the lower classes, but their histories also do reveal to some extent the terror and confusion among the lower classes in regard to both the disease and state measures against it. See also La Berge, *Mission and Method.* La Berge argues that the failure of the early French public health movement was due to the liberal government in France at the time, whereas in Britain the more interventionist government allowed public health to move forward as a field (p. 177). She also, interestingly, refers to hygienism as "medical imperialism" in the introduction. In contrast, Ackerman examines not the state interventions, but the very slow acceptance of "modern medicine" by peasants in the Parisian countryside in the nineteenth and early twentieth century. Evelyn Ackerman, *Health Care in the Parisian Countryside, 1800–1914* (New Brunswick: Rutgers University Press, 1990).
16. See Anthony Reid, *Southeast Asia in the Age of Commerce: 1450–1680, Volume One: The Lands Below the Winds* (New Haven: Yale University Press, 1988); O. W. Wolters, *History, Culture and Region in the Southeast Asian Perspectives* (Singapore: Institute of Southeast Asian Studies, 1982). Wolters argues that the "closed sea" aspect of this open trade area is a possible shared shaper of Southeast Asian cultures.
17. For review of major economic trends in Southeast Asia in the nineteenth century, see Nicholas Tarling, ed., *The Cambridge History of Southeast Asia: Volume 2 Part One: From C.1800 to the 1930s* (Cambridge: Cambridge University Press, 1992).
18. Giovanni Berlinguer, "Globalization and Global Health," *International Journal of Health Services* 29, no. 3 (1999).
19. Obligatory declaration denotes the required reporting of sick individuals to government authorities.
20. NAC RSC 7878. The numbering of the diseases was consistent through the colonial period. The 1911 circular fixing diseases for which declaration was obligatory lists

(1) typhoid fever and paratyphoid fevers, (2) exanthemic typhoid, (3) smallpox, (4) scarlet fever, (5) diphtheria, (6) military fever (*suette miliaire*, sometimes *suette militaire*), (7) cholera, (8) dysentery, (9) the plague, (10) yellow fever, (11) puerperal fever, (12) newborn ophthalmia, (13) measles, (14) Malta fever, (15) leprosy, (16) recurrent fever, (17) cerebro-spinal meningitis, (18) acute pyroplasmosis or tropical splenomegalia.

21. As discussed in previous chapters, "*dépistage*, containment, and disinfection" could entail a variety of measures including forced isolation in lazarets, vaccinations, home quarantines, sanitary passports, and property incineration. With the exception of the sanitary passport, these measures will be described in other chapters in detail. The sanitary passport was a document given to neighbors and family of epidemic disease victims. Holders of the passport had to show up for a doctor's examination for ten consecutive mornings before they were allowed freedom of movement. These passports did not seem to be very effective, although they were used through much of the colonial period. See NAC RSC 2485.
22. NAC RSC 7878. See also chapter 2.
23. France was not alone in its research; all of the colonial powers in Southeast Asia focused scientific attention on the disease. NAC RSC 1357, 1358, 18241.
24. Quinine is still an effective treatment against malaria. Its major problems are negative side effects and an unpleasant, intensely bitter flavor.
25. NAC RSC 2486.
26. NAC RSC 12620.
27. The *mesrok* is the indigenous provincial chief.
28. See NAC RSC 12620, 25666, and 28758. The AM enacted a similar countrywide program of stovarsol distribution for yaws in the 1930s.
29. Ronald Ross published his findings on the role of the anopheles mosquito in the spread of malaria in the *British Medical Journal* in 1897.
30. NAC RSC 1248.
31. For an extensive report on malaria and engineering works in Vietnam, see NAC RSC 605, 1160, 1357 and 1358. File 605 contains a report on the introduction of foreign fish to eat the anopheles mosquito larvae. The report cites the disastrous economic effect of such efforts in other colonies, reducing native fish stock and profits from fishing. File 1160 contains the report of the Pasteurian Morin as part of the PI Antimalarial Service in 1931. File 1357 contains lengthy reports from the PI and the Public Works Department on malaria eradication schemes. For the most part these schemes called for creating brush-free zones, encouraging native cleanliness, and killing mosquito larvae.
32. NAC RSC 1200.
33. Ibid. His observation is odd, since quinine is still today an effective last-resort medicine against malaria. Also, quinacrine's major side effect—its tendency to turn long-term users yellow—ended up limiting its use.
34. Robert A. Nye, *Crime, Madness, and Politics in Modern France: The Medical Concept of National Decline* (Princeton, NJ: Princeton University Press, 1984).
35. Ibid. p. xii.
36. Eugen Weber, *The Hollow Years: France in the 1930s* (New York: W. W. Norton and Company, 1996), chapter 2.
37. See William Schneider, *Quality and Quantity: The Quest for Biological Regeneration in Twentieth-Century France* (New York: Cambridge University Press, 1990), introduction. France's lack of such negative eugenics laws is in contrast to practices in

countries such as the United States and Germany, which legalized sterilization of the feeble-minded and insane. See Daniel J. Kevles, *In the Name of Eugenics: Genetics and the Uses of Human Heredity*, 2nd ed. (Cambridge, MA: Harvard University Press, 1995).

38. Because of this strong link with the natalist movement, I will reserve further discussion of eugenics and depopulation to the chapter on women's and children's health.
39. NAC RSC 605. Letter to the GGI, April 6, 1938.
40. Ibid.
41. One historian argued that for subjects in French colonial Africa, the naturalization laws passed in these decades were designed to encourage rapid inclusion of second-generation immigrants as French citizens but the exclusion of the colonial subject. Alice Conklin, *A Mission to Civilize: The Republican Idea of Empire in France and West Africa, 1895–1930* (Stanford, CA: Stanford University Press, 1997). Her analysis does not extend to a reading of these measures in other French colonies. With the partial exception of one essay by Ann Stoler, a study considering the impact in colonial Indochina of race, science, and law has yet to be written. See Ann Laura Stoler, "Sexual Affronts and Racial Frontiers," in *Tensions of Empire: Colonial Cultures in a Bourgeois World*, ed. Ann Laura Stoler and Frederick Cooper (Berkeley: University of California Press, 1997).
42. For a concise political analysis of the Popular Front, see Weber, *The Hollow Years: France in the 1930s*, chapter 6. See also Gordon Wright, *France in Modern Times*, 5th ed. (New York: W. W. Norton, 1995), chapters 30 and 31.
43. NAC RSC 2192.
44. The malarial index or "*index paludéen*" was created by taking and examining blood samples of large segments of the population, thereby estimating what percentage of the population was actually afflicted by the disease. These records may be of use to those studying long-term epidemiological health records for the region. Good records of malarial indexes exist for the Haut Chhlong region in the late 1930s.
45. NAC RSC 1248. Letter from GGI to regional heads, July 5, 1922, clarifying the purpose of these sectors.
46. AC RSC 1248.
47. Menaut's 1923 annual report for the experimental hygiene sector provides a fascinating glimpse of the village life of subsistence farmers. The resounding impression it gave was that hygiene would improve when poverty was ameliorated and material well-being increased. Menaut also created a detailed survey on potable water in the region, carefully collecting information on wells, land elevation, and needed improvements. NAC RSC 1248.
48. NAC RSC 1355, 2187.
49. NAC RSC 5077.
50. NAC RSC 4557.
51. NAC RSC 4958.
52. NAC RSC 1140. Menaut would serve off and on as health director from 1927 to his departure from Cambodia in 1939, usually in the role of an interim director between permanent directors. It seemed to be his choice not to remain in the position. He died in 1943 in Rion-des-Landes, France, of an unspecified liver ailment he had contracted in Cambodia.
53. See Martin Bernal, "The Nghe-Tinh Soviet Movement 1930–1931," *Past and Present*, no. 92 (1981); Milton Osborne, "Continuity and Motivation in the Vietnamese

Revolution: New Light from the 1930's," *Pacific Affairs* 47, no. 1 (1974). For the repression that followed the movement, see Peter Zinoman, *The Colonial Bastille: A History of Imprisonment in Vietnam, 1862–1940* (Berkeley: University of California Press, 2001), p. 110. See also chapter 6 of the present study.

54. James C. Scott, *The Moral Economy of the Peasant: Rebellion and Subsistence in Southeast Asia* (New Haven: Yale University Press, 1976), p. 120. See also Bernal, "The Nghe-Tinh Soviet Movement 1930–1931"; Osborne, "Continuity and Motivation."
55. NAC RSC 2391. Circular no. 6, 1930, from the inspector general to regional directors.
56. Ibid. Response of Health Director Menaut to the inspector general, March 3, 1930.
57. Hy Van Luong makes a similar point regarding the assumed relationship between the Great Depression and famine of 1930 and the Nghe-Tinh soviet movement. He argues against several scholars of peasant revolution (including James Scott and Jeffrey Paige) that economic hardship was not the ultimate cause of the 1930 uprising, although it may have played a minor part. Hy Van Luong, "Agrarian Unrest from an Anthropological Perspective: The Case of Vietnam," *Comparative Politics* 17, no. 2 (1985).
58. This questionnaire also points to an awareness, on some level, of social problems concomitant with economic and physical hardship *before* any serious social unrest—an awareness that many scholars claim was totally lacking in the colonial state. However, some analyses of the Nghe-Tinh movement point to French knowledge of the unrest in the area well before the rebellion began.
59. NAC RSC 902, 2386.
60. NAC RSC 2192.
61. NAC RSC 2356. Letter from Lavit to GGI, undated. Indicative of a renewed thrust towards public health in 1931, the AM Annual Medical Report (Assistance Médicale rapport annual du Cambodge) was renamed Report on the Functioning of the Sanitary and Medical Services of Cambodia (Rapport sur le fonctionnement des services Sanitaires et Médicaux du Cambodge). NAC RSC 1675.
62. NAC RSC 902.
63. NAC RSC 31820.
64. Ibid. This circular, while not legislating changes, urged medical care providers to shift their focus away from acute care at the hospitals and towards general care of rural populations and disseminating advice and information.
65. NAC RSC 7637.
66. NAC newspaper collections. *Echo du Cambodge*, March 16, 1938.
67. Daniel Panzac, *Quarantaines et lazarets: L'Europe et la peste d'Orient* (Aix-en-Provence: Édisud, 1986).
68. See Barnes, *The Making of a Social Disease*, René Dubos, *Louis Pasteur, Free Lance of Science*, 2nd ed. (New York: Da Capo Press, 1960). Dubos points out that the debate of soil versus seed originates in the Hippocratic era with the debate between cos (soil/patient) and cnidia (seed/disease), p. 272. Gerald L. Geison, *The Private Science of Louis Pasteur* (Princeton, NJ: Princeton University Press, 1995); Bruno Latour, *The Pasteurization of France*, trans. Alan Sheridan and John Law (Cambridge, MA: Harvard University Press, 1988); Jacques Léonard, *La médecine entre les pouvoirs et les savoirs* (Paris: Aubier Montaigne, 1981), especially pp. 156–59 on hygienists' alignment with the changing ideas and growing importance of the new concept of contagion. Hildreth argues that there was intensified emphasis on the family as a medical unit at

the end of the nineteenth century because of the growing fear of medicine becoming too impersonal, reductive, or mechanical. Martha L. Hildreth, "Doctors and Families in France, 1880–1930: The Cultural Reconstruction of Medicine," in *French Medical Culture in the Nineteenth Century*, ed. Ann La Berge and Mordechai Feingold (Atlanta: Editions Rodopi, 1994).

69. For France, this is shown by Barnes, *The Making of a Social Disease*. To a lesser extent it is also shown in Aisenberg, *Contagion*. In *The White Plague*, the Duboses show how, with the rise in germ theory, public understanding of tuberculosis in Europe shifted from the concept of a romantic disease to that of a dirty disease. René Dubos and Jean Dubos, *The White Plague: Tuberculosis, Man, Society* (Boston: Little, Brown and Company, 1952). For America, one of the best studies of the adoption of germ theory in the domestic sphere is Nancy Tomes, *The Gospel of Germs: Men, Women and the Microbe in American Life* (Cambridge, MA: Harvard University Press, 1998). Hammonds' study includes detailed analysis of public education campaigns on the diphtheria "germ" in New York. Evelynn Maxine Hammonds, *Childhood's Deadly Scourge: The Campaign to Control Diphtheria in New York City 1880–1930* (Baltimore: Johns Hopkins University Press, 1999).
70. *Chhlong* is also important in another medical sense. The compound phrase *chhlong tonlé* (literally "cross the lake") refers to giving birth. There are many other compound words incorporating *chhlong*.
71. Erwin H. Ackerknecht, "Anticontagionism between 1821 and 1867," *Bulletin of the History of Medicine* 22, no. 5 (1948).
72. Epidemic severity is often determined by the population density in an area. The growth of larger towns and cities with French colonialism meant an increase of population density and, likely, epidemic severity.
73. Again, this is interesting to compare with French views in the seventeenth and eighteenth centuries, when odors were considered agents of disease. See Alain Corbin, *The Foul and the Fragrant: Odor and the French Social Imagination* (Cambridge, MA: Harvard University Press, 1986).
74. NAC RSC 12618.
75. NAC RSC 2140.
76. Today, the Khmer term is *merook*.
77. NAC RSC 26705.
78. See the story of Yersin's plague research in chapter 2.
79. NAC RSC 25666.
80. NAC RSC 1248.
81. Conjunctivitis was a continual problem in the Khmer countryside. In March 1920, the *resident* of Stung Treng had distributed a poster in Khmer attempting to explain the causes of the disease. A copy of the announcement does not exist, but it seems to have been ineffective. NAC RSC 25607.
82. NAC RSC 905. The Khmer phrase for disgusting is *khpum*.
83. If this poster changed local behavior, the change was temporary.
84. A *mekhum* is the chief of a *khum*.
85. NAC RSC 5127.
86. In fact, due to local etiology, malaria was a later source of linguistic confusion for French researchers. The Khmer term for malaria is *kron chanh* (*kron* = fever and *chanh* = to lose, fail, or be defeated; thus literally, the "losing/failing fever"); however, its full terminology is *kron chanh-dey* (awkwardly translated as "fever of being defeated by

the earth"). In 1939, the French encountered an epidemic of a disease asymptomatic of malaria in a sparsely populated region that the administration was attempting to develop for plantation work. Confusingly, natives also referred to the disease as *kron chanh,* but more specifically as *kron chanh-tuk* ("fever of being defeated by water"). NAC RSC 605. The health services dispatched a group of doctors to study the disease. One of them was himself afflicted by the mystery illness during the course of his study, and was strongly of the opinion that it was distinct from malaria. Other researchers thought perhaps it was a secondary infection brought on by the disease, or a related form of infection. Unable to come to a diagnosis of a specific disease, much less its etiology, the researchers could make few guesses on possible preventative measures. All authors urged further analysis; however, the coming of World War II and its aftermath made further study moot in terms of French economic development. It might be that this was simply one of the less common forms of the malaria plasmodium, but this disease remains a historical medical mystery.

87. NAC RSC 11373, 27522.
88. NAC RSC 686.
89. NAC RSC 99. GGI circular no. 64, April 14, 1925.
90. These four schools were the College Sisowath, École François Baudoin, École Chak Angre, and the École Pheak Buon. NAC RSC 26703.
91. NAC RSC 12618.
92. NAC RSC 2273. These elementary rules included boiling all drinking water and cooking all vegetables before eating them, particularly at the end of the rainy season.
93. Ibid. The proliferation of honorific titles was quite fashionable in this period.
94. Ernest Gellner, *Nations and Nationalism* (Ithaca, NY: Cornell University Press, 1983), p. 21.
95. NAC RSC 116. Pursat, 1927. The report was written in French.
96. Edwards, *Cambodge: The Cultivation of a Nation, 1860–1945,* p. 228.
97. Stacy Leigh Pigg, "Acronyms and Effacement: Traditional Medical Practitioners (Tmp) in International Health Development," *Social Science & Medicine* 41, no. 1 (1995);and "Found in Most Traditional Societies: Traditional Medical Practitioners between Culture and Development," in *International Development and the Social Sciences,* ed. Frederick Cooper and Randall Packard (Berkeley: University of California Press, 1997).
98. Dr. Adrien Pannetier, *Au coeur du pays khmer: Notes cambodgiennes,* 2nd ed. (Paris: CEDORECK, 1983).
99. NAC RSC 102.
100. NAC RSC 12635.
101. NAC RSC 2414.
102. It would seem that the appeal to French courts to hear magical cases was not uncommon. This ceremony was similar to other, more recent accounts of magical healing ceremonies in Cambodian *wats*. See chapter 2.
103. Anne Ruth Hansen, *How to Behave: Buddhism and Modernity in Colonial Cambodia, 1860–1930* (Honolulu: University of Hawaii Press, 2007), pp. 58–60.
104. NAC RSC 23452, 28874, 28944. In one of the most interesting Moi investigations, a woman had two children (a boy and a girl) with an involved and fascinating mythology surrounding both their births and their magical attributes. They also were possessed of all-healing lustral water, and apparently had amassed a great deal of tributary wealth from surrounding villagers. CAOM INDO RSC collections, file 235.

105. Candle wax is a religiously significant gift often given to monks. The doctor estimated at the time of his visit that the young boy had accrued about sixty kilograms of wax, as well as a large pile of incense sticks.
106. NAC RSC 2414.
107. For the story of the development of the yaws elimination campaign, see NAC RSC 134, 605, 2115, 2118, 4958, and 8841.

CHAPTER FIVE

1. These charts are compiled with data from district annual reports. Because the data was often changed, omitted, or condensed in the ensemble report for the country, I chose two representative districts with relatively complete numbers for male, female, and infant consultations. While medical interest in women was highest in the second half of the 1930s, this ironically was also when most district medical reports dropped tallies by sex.
2. Alisa Klaus, *Every Child a Lion: The Origins of Maternal and Infant Health Policy in the United States and France, 1890–1920* (Ithaca, NY: Cornell University Press, 1993), see the introduction (particularly p. 6) and chapters 1 and 2.
3. Margaret Jolly, "Colonial and Postcolonial Plots in the History of Maternities and Modernities," in *Maternities and Modernities: Colonial and Postcolonial Experiences in Asia and the Pacific,* ed. Margaret Jolly and Kalpana Ram (Cambridge: Cambridge University Press, 1998), p. 2.
4. Seth Koven and Sonya Michel, "Womanly Duties: Maternalist Politics and the Origins of Welfare States in France, Germany, Great Britain, and the United States, 1880–1920," *American Historical Review* 95, no. 4 (1990), p. 1079.
5. Jane Haggis, "'Good Wives and Mothers' or 'Dedicated Workers'?" in *Maternities and Modernities: Colonial and Postcolonial Experiences in Asia and the Pacific,* ed. Margaret Jolly and Kalpana Ram (Cambridge: Cambridge University Press, 1998).
6. Barbara Andaya, "Introduction," in *Other Pasts: Women, Gender and History in Early Modern Southeast Asia,* ed. Barbara Andaya (Manoa: University of Hawaii Press, 2000), p. 6. Andaya also remarks that "any comments regarding [the Southeast Asian woman's] purported status must be considered contextually." p. 4.
7. For a discussion of the high status of women in Southeast Asian society, see, for example, the essays in Penny Van Esterik, ed., *Women of Southeast Asia* (DeKalb, IL: Northern Illinois University, Center for Southeast Asian Studies, 1982). Southeast Asian historians frequently claim the uniquely high social status of women as one of the features distinguishing Southeast Asia as a coherent region.
8. Marie-Paule Ha, "Engendering French Colonial History: The Case of Indochina," in *The French and the Pacific World, 17th–19th Centuries: Explorations, Migrations and Cultural Exchanges,* ed. Annick Foucrier (Burlington, VT: Ashgate, 2005), p. 166.
9. Scholars have increasingly addressed the experiences of European women in the colony, and yet the lived experiences of indigenous women remain somewhat elusive. Jean Gelman Taylor's now classic *The Social World of Batavia: European and Eurasian in Dutch Asia* (Madison: University of Wisconsin Press, 1983) is an excellent history of European and mixed women in the colony. We have nothing comparable for indigenous women.
10. Philippa Levine, "Rereading the 1890s: Venereal Disease As 'Constitutional Crisis' in Britain and British India," *The Journal of Asian Studies* 55, no. 3 (1996): p. 587.
11. Penny Edwards, "'Propagender': Marianne, Joan of Arc and the Export of French Gender Ideology to Colonial Cambodia (1863–1954)," in *Promoting the Colonial Idea:*

*Propaganda and Visions of Empire in France*, ed. Tony Chafer and Amanda Sackur (New York: Palgrave Publishers, 2002), pp. 116–17. Edwards contends that in precolonial Cambodia, "gender equity was subtly reinforced by strong similarities in traditional clothes and the coiffures of men and women." p. 117. The reverse could easily be argued with the same logic. That is, such similarities in appearance are feasible *because of* the stark difference in the social role of men and women, which certainly does not preclude gender inequality. Overt markers are less necessary if social roles are strongly constrained in other ways. The many different rituals surrounding the maturation of adolescent Khmer males and females are an obvious example of the creation of divergent masculinity and femininity.

12. NAC RSC 9614 (Maurel's journal of his 1885 efforts), and NAC RSC 11197 (Hahn's 1886 report on Maurel's service).
13. NAC RSC 33882.
14. Gregor Muller, *Colonial Cambodia's "Bad Frenchmen": The Rise of French Rule and the Life of Thomas Caraman, 1840–1887* (New York: Routledge, 2006), pp. 138–46.
15. For the history of prostitution in France in the nineteenth and twentieth centuries, see Alain Corbin, *Women for Hire: Prostitution and Sexuality in France after 1850*, trans. Alan Sheridan (Cambridge, MA: Harvard University Press, 1990).
16. NAC RSC 33882.
17. One scholar incorrectly identifies this as the first general dispensary for all Cambodian women.
18. NAC RSC 2025, 12632, 33882. NAC newspaper collections. CAOM Fonds Indochine RSC 278.
19. NAC RSC newspaper collection. *Le Populaire*, June 20, June 21, and July, 2, 1935. These articles provide a vivid and sympathetic account of the events precipitating the riot.
20. NAC RSC 78. This may be akin to the phenomena described by Aihwa Ong in her study of Malaysian factory workers. Aihwa Ong, "The Production of Possession: Spirits and the Multinational Corporation in Malaysia," in *Women, Gender, Religion: A Reader*, ed. Elizabeth Castelli (New York: Palgrave Press, 2001).
21. John F. Cady, *The Roots of French Imperialism in Eastern Asia* (Ithaca, NY: Cornell University Press, 1954).
22. François Ponchaud, *La cathédrale de la rizière: 450 ans d'histoire de l'église au Cambodge* (Paris: Librarie Arthème Fayard, 1990).
23. Before the institution of the AM, two to three sisters also staffed the Phnom Penh hospital. However, the relationship between the Catholic Church and the AM was rocky, and the sisters were quickly dropped from the hospital staff. NAC RSC 11199.
24. NAC RSC 32861. The ratio of Vietnamese to Khmer in the hospital and crèche of the Soeurs de la Providence remained approximately nine to one through the colonial period.
25. Khmers are predominantly Theravada Buddhists.
26. It is sometimes spelled Cualao Gieng. The town is close to Saigon.
27. Abbé Grandjou, *La congrégation des Soeurs de la Providence de Portieux (Vosges) fondée par le vénérable Jean Martin Moyë en 1762* (Lille: Desclée, de Brouwer & Co., 1923).
28. NAC RSC 2190.
29. NAC RSC 12636.
30. Ibid.
31. Ibid.

32. This seemed to be a programmatic commitment of the entire order (which had branches in France, Belgium, Rome, and Manchuria as well as Indochina). The Order of the Sisters of Providence had been founded by a Father Möye in the late eighteenth century after his missionary experience in China, where he had "found" three thousand dying infants whom he had successfully baptized as "a veritable army of angels, who will eternally sing the glory of the Lord in reveling in his infinite joy." Möye's explicit purpose in forming the order was to continue this work of making these infant angels. Grandjou, *La congrégation des soeurs*, p. 15.
33. NAC RSC 1149.
34. NAC RSC 2117.
35. The fraught relationship between republican freemasons and Catholic missionaries in the colonies has been well researched in a recent dissertation. J. P. Daughton, "The Civilizing Mission: Missionaries, Colonialist, and French Identity 1885–1914" (PhD dissertation, University of California, Berkeley, 2002). The vast ideological gap between anticlerical intellectuals and Catholic traditionalists in France has also been well analyzed, most notably in scholarship surrounding the Dreyfus affair. See, for example, Jean-Denis Bredin, *The Affair: The Case of Alfred Dreyfus*, trans. Jeffrey Mehlman (New York: George Braziller, 1983).
36. NAC RSC 25715.
37. NAC RSC 2454. The first French colonists in Indochina, in the last quarter of the nineteenth century, were generally much more willing to recognize their offspring than the colonists of later generations. It was just at the turn of the century when unclaimed *métis* were increasingly perceived as a problem, as the date of the founding of the SPE indicates. For more on the earlier colonists and their relationship with their mixed children, see chapter 5 of Muller, *Colonial Cambodia's "Bad Frenchmen."*
38. In Hanoi, for example, the Société d'Assistance aux Enfants Franco-Indochinois began in 1898 as the Société des Métis, and changed its name in 1925. NAC RSC 2287. For a chronological list of *métis*-sponsoring organizations in French Indochina, see J. Mazet, *La condition juridique des métis dans les possessions françaises* (Paris: Editions Domat-Montchrestien, 1932).
39. NAC RSC 2454.
40. Despite the wealth of archival materials, a detailed history of the *métis* population in Indochina has yet to be written. Ann Stoler has written two articles concerning the *métis*. Ann Laura Stoler, "Sexual Affronts and Racial Frontiers," in *Tensions of Empire: Colonial Cultures in a Bourgeois World*, ed. Ann Laura Stoler and Frederick Cooper (Berkeley: University of California Press, 1997); and "Rethinking Colonial Categories: European Communities and the Boundaries of Rule," in *Colonialism and Culture*, ed. Nicholas B. Dirks (Ann Arbor: University of Michigan Press, 1992). A historical study of *métis* is partly complicated by their lack of a legal status. See Pierre Guillaume, "Les métis en Indochine," *Annales de démographie historique* (1995). Guillaume is forced to estimate the number of *métis* indirectly from other types of records—for example, of marriages of French men with indigenous women. The fragmentary records from the NAC files of Cambodia's *métis* "orphans" indicates that some of them kept in contact with their indigenous mothers after becoming wards of the SPE. This would strongly suggest that certain of these children were often not abandoned so much as relinquished.
41. While organizations such as the SPE were explicit in linking their works to the maintenance of French prestige, others also identified the *métis* with potential political

unrest. One military commander would comment that left to their own devices, they "constitute[d] a danger if left to be exploited by malcontents." NAC RSC 2452.

42. NAC RSC 1315.
43. In 1911, Dr. Dubalen of Svay Rieng suggested an even larger bribe of four to five piastres for giving birth in maternities. NAC RSC 15085. A *layette* is an outfit and bedding set for newborns.
44. NAC RSC 26711.
45. Ibid.
46. The riel is the Khmer unit of currency.
47. NAC RSC 26711. Whether the AM simply did not concern itself with or did not catch this specification is unclear.
48. NAC RSC 709.
49. Eugene Declerq et al., "Where to Give Birth," in *Birth by Design: Pregnancy, Maternity Care, and Midwifery in North America and Europe*, ed. Raymond De Vries (New York: Routledge, 2001). See also Ann Oakley, *The Captured Womb: A History of the Medical Care of Pregnant Women* (New York: Blackwell Publishers, 1984), p. 132. Sixty percent of births in Britain immediately before World War II were still in the home.
50. NAC RSC 12634.
51. The SPMI eventually opened a small consultation office in a building separate from the mayor's office.
52. NAC RSC 2117.
53. NAC RSC 88.
54. NAC RSC 709 and 7637.
55. NAC RSC 26681.
56. NAC RSC 31889.
57. For a more detailed analysis of this phenomenon, see chapter 4.
58. William Schneider, *Quality and Quantity: The Quest for Biological Regeneration in Twentieth-Century France* (New York: Cambridge University Press, 1990).
59. Ibid. See chapters 3 and 4.
60. Anne Witz, "'Colonising Women': Female Medical Practice in Colonial India, 1880–1890," in *Women and Modern Medicine*, ed. Lawrence Conrad and Anne Hardy (Amsterdam: Editions Rodopi B. V., 2001). Also see the introduction of Jonathan Glustrom Katz, *Murder in Marrakesh: Émile Mauchamp and the French Colonial Adventure* (Bloomington: Indiana University Press, 2006).
61. Etienne Aymonier, *Notes sur les coutumes et croyances superstitieuses des cambodgiens* (Paris: CEDORECK, 1984 [1883]), p.73.
62. See chapter 3.
63. As mentioned in chapter 3, the *priey kraalaa pleung* are the spirits of the unborn fetus and the mother who died in childbirth, generally seen as the angriest and most destructive in the pantheon of ghosts.
64. The rites and precautions surrounding rural birth in Cambodia seem not to have changed appreciably from the writing of late-nineteenth-century scholars to the writings of late-twentieth-century WHO researchers.
65. Ang Choulean, "Grossesse et accouchement au Cambodge: Aspects rituals," *ASEMI* XIII (1982), p. 92.
66. Ibid., p. 97.
67. Among some ethnic minorities in Cambodia, the father could be called upon to ritually protect the outside perimeter of the birthing hut. Jacqueline Matras-Troubetzkoy,

*Un village en forêt: L'essartage chez les Brou du Cambodge* (Paris: SELAF, 1983), p. 95.

68. See Choulean, "Grossesse et accouchement," passim; Patrice M. White, "Crossing the River: Traditional Beliefs and Practices of Khmer Women during Pregnancy, Birth and Postpartum," (Phnom Penh: PACT/JSI and Cambodian Ministry of Health, 1995); Ritharasi Norodom, "L'évolution de la médecine au Cambodge" (M.D. thesis, Librarie Louis Arnett, 1929). Roasting time varies greatly, but on average is two or three days. The amount of time allotted to this practice does not seem to have appreciably changed over the years. For instance, White's 1995 report does not contradict any of Poree-Maspero's information from 1958. See Evelyn Porée-Maspero, *Cérémonies privées des cambodgiens* (Phnom Penh: Éditions de L'Institut Bouddhique, 1958). This is also not a practice exclusive to Cambodia. See Lenore Manderson, "Roasting, Smoking and Dieting in Response to Birth: Malay Confinement in Cross-Cultural Perspective," *Social Science & Medicine* 15B, no. 4 (1981).
69. Ang Choulean, "De la naissance à la puberté. Rites et croyances khmers," in *Enfants et sociétés d'Asie du Sud-Est*, ed. Jeannine Koubi and Josiane Massard-Vincent (Paris: Harmattan, 1994), p. 156.
70. NAC RSC 1729.
71. NAC RSC 1814.
72. NAC RSC 1315.
73. NAC RSC 4554.
74. NAC RSC 108.
75. NAC RSC 78.
76. NAC RSC 29641. Letter from Simon to RSC, June 15, 1936. Capital letters in original.
77. NAC RSC 7637.
78. NAC RSC 32225.
79. NAC RSC 2486. AM annual report 1908. DLS Hauer. See also Sokhieng Au, "Indigenous Politics, Public Health and the Cambodian Colonial State," *Southeast Asia Research* 14, no. 1 (2006).
80. NAC RSC 2140. Letter from Mayor Leclère to RSC, May 25, 1908.
81. NAC RSC 11082.
82. NAC RSC 97.
83. NAC RSC 2479.
84. NAC RSC 1675. Health Director annual report, 1930.
85. NAC RSC 134. Takeo annual medical report, 1931.
86. NAC RSC 26681.
87. NAC RSC 8843.
88. It is inaccurate to call the SPE or SPMI entirely private initiatives, since so many prominent members of colonial government were associated with these organizations, either directly or through their wives.
89. NAC RSC 26695. Battambang annual report 1934.
90. NAC newspaper collections. *Echo du Cambodge*, May 19, 1936.
91. NAC RSC 2455.
92. Some scholarship addresses the influence of empire on metropolitan eugenics. In a book edited by Ann Stoler and Frederick Cooper, one author discusses how possession of empire encouraged eugenicists' rhetoric in prodding British mothers to reproduce and "fill the empty spaces of empire" (p. 88). Anna Davin, "Imperialism

and Motherhood," in *Tensions of Empire: Colonial Cultures in a Bourgeois World*, ed. Ann Laura Stoler and Frederick Cooper (Berkeley: University of California Press, 1997). Little scholarship has been produced analyzing the manifestations of an "imperial" eugenics in the colonies. Nancy Rose Hunt, in another chapter in *Tensions of Empire*, discusses the intervention of the colonial regime of the Belgian Congo in maternal and infant health. She never explicitly mentions eugenics, but does point to the growing colonial concern about population loss in the 1920s. The programs that she describes in the Congo are similar to those implemented in Cambodia—for example, mothers receiving food, soap, clothes, and money for giving birth at the hospital or attending post-natal consultations (p. 298). Nancy Rose Hunt, "Le Bébé en Brousse: European Women, African Birth Spacing, and Colonial Intervention in Breast Feeding in the Belgian Congo," in *Tensions of Empire: Colonial Cultures in a Bourgeois World*, ed. Ann Laura Stoler and Frederick Cooper (Berkeley: University of California Press, 1997). To my knowledge, nothing has been written of this in Southeast Asia. However, other colonial governments in Southeast Asia clearly partook of the eugenic thinking that was in vogue in Europe and America. For example, Judith Richell incidentally mentions widespread "baby shows" in Burma during the interwar period, although she labels them as part of the effort to reduce infant mortality (which, as I have been arguing above, was a eugenic goal in the French context). Judith Richell, "Ephemeral Lives: The Unremitting Infant Mortality of Colonial Burma, 1891–1941," in *Women and Children First: International Maternal and Infant Welfare, 1870–1945*, ed. Valerie Fildes, Lara Marks, and Hilary Marland (New York: Routledge, 1992).

93. NAC newspaper collections. *Echo du Cambodge*, September 10, 1935.
94. Barbara Andaya, "Localising the Universal: Women, Motherhood and the Appeal of Early Theravada Buddhism," *Journal of Southeast Asian Studies* 33, no. 1 (2002): p. 21.
95. Ibid.
96. NAC newspaper collections. *La Presse Indochinoise*, September 13, 1935.
97. NAC newspaper collections. *La Presse Indochinoise*, August 24, 1935.
98. NAC RSC 7638. The annual report for 1939 lists a maternity for each of the fourteen districts outside of Cambodia, yet, for most of these districts, this maternity was simply a room, or part of a room, reserved for giving birth. See NAC RSC 2118.
99. NAC RSC annual medical reports.
100. NAC RSC 8841.
101. NAC RSC 26701.
102. NAC 2118.
103. See chapter 3 for a history of men's education at the École.
104. Please see explanatory note in apage on abbreviations.
105. NAC RSC 1315.
106. NAC RSC 2246.
107. NAC RSC 20109.
108. NAC RSC 14640.
109. NAC RSC 1148 and 1249.
110. NAC RSC 1249.
111. Ibid. See also chapter 3.
112. NAC 14640.
113. NAC RSC 27018.
114. NAC RSC 27018.

115. Of the eight entrants in year one, four did not complete the two-year training program. Of the four who did finish, one (Saroun) attempted to resign immediately but could not pay back the training fees. In the second year, only five students entered. One dropped out immediately. Two others likely dropped out, as no records exist for them except their entry notifications. Of the five students in the third year, three did not complete the program, and we have no record of what happened to a fourth (Pauline Chea).
116. NAC RSC 11288.
117. NAC RSC 14640.
118. NAC RSC 3009. A detailed description of one exam can be found in NAC 2237.
119. See, for example, NAC RSC 7637.
120. NAC RSC 11996.
121. NAC RSC 3011
122. NAC RSC 11996.
123. NAC RSC 7185. Although her aborted career is not mentioned, Tit Saom's bureaucratic requirements upon entering into the program are detailed in Kate Frieson, "Sentimental Education: Les Sages Femmes and Colonial Cambodia," *Journal of Colonialism and Colonial History* 1, no. 1 (2000). This article also argues the constrained public role of Khmer women in relation to men. Please note that the claim of a Cambodian Maternity School in 1912 is incorrect, stemming from an incorrect translation of École Maternelle (nursery school) from NAC RSC 417. Another incorrect claim in the article also stems from a mistranslation: the women's dispensary built in 1922 was not a public dispensary for girls, but a dispensary for "public" girls (prostitutes), administered in conjunction to the prison system (see beginning of this chapter).
124. NAC RSC 11996. Letter from DLS Simon to RSC, July 15, 1938.
125. NAC RSC 10934. See also chapter 3.
126. Ibid.
127. Ibid. Letter from Nguyen Thi Lieng's aunt to the École de Médecine.
128. Ibid. Letter from the EM to the Cambodian DLS, February 7, 1911.
129. NAC RSC 10934, 25675.
130. NAC RSC 10803.
131. NAC RSC 27018, 6848, 11326. These include Madeleine Ba (Humpbert), Nguyen Thi Hoa (Marguerite Rigaud), and Nguyen Thi Ngoc, all in the photo gallery (figure 5.9). See also NAC RSC 10644 (Nguyen Thi Dung), 6347 (Le Kim Nang), and 7218 (Nguyen thi Hieu, or Jeanne Nguyen), from other cadres, who resigned for health reasons.
132. NAC RSC 7218.
133. NAC RSC 8325, 1148. Sage-femme work may have become a family affair for the Reynauds. In the postcolonial period, Josephine Reynaud opened a private maternity in Phnom Penh in 1955. Cambodian Government, *Journal Officiel Du Cambodge* (Phnom Penh: Cambodian Government, 1955).
134. NAC RSC 27018. In another example, a young sage-femme trained in Vietnam, Nguyen Van Sanh, was initially posted in Phnom Penh in 1933. The colonial government posted her Vietnamese husband in Saigon. Thus, she requested relocation to Svay Rieng, which was marginally closer to Saigon than Phnom Penh. The AM denied her request. From 1934 onwards, she requested short stints of unpaid sick live, which seemed to be an effective ploy to delay repayment of school fees. She finally left in 1936 for three years of unpaid sick leave. However, when she returned in late

1939, she left unannounced shortly thereafter, and the administration was forced to revoke her employment. NAC RSC 10580.

135. NAC RSC 7073, 11292. Ngo Thi Phuoc was technically fired when she was reposted but did not report to Phnom Penh after several years serving in Banam, where her husband and children lived.
136. Nguyen Thi Yen, a Vietnamese woman trained at the EM midwife program, began her AM career on shaky footing, with her supervisor noting that she was intelligent but "difficult." In 1925 she requested a posting in Phnom Penh, but was assigned to Prey Veng, where she grudgingly reported. In her first six months she seemed to have adjusted, receiving a very good performance review by her supervisor. However, the health director then reassigned her to Kompong Chhnang because the Prey Veng maternity had had only one birth in that six-month period. She promptly resigned. NAC RSC 7073.
137. At that time the two cities were probably at minimum a full day's journey apart for anyone who lacked a motorboat.
138. NAC RSC 7187.
139. NAC RSC 7187.
140. Ibid. Letter to the RSC from Neang Saroun in Siem Reap, May 20, 1930.
141. Ibid.
142. NAC RSC 2452.
143. NAC RSC 2452. Quoting a January 1915 *Revue Indochinoise* interview.
144. Ibid.
145. The métis children often took the name of their father. See for example NAC RSC 6848, on Marguerite Rigaud, who was never recognized by her father. She entered the midwife program in the late 1920s. We do not know if this is the same Rigaud mentioned in chapter 3.
146. NAC RSC 6415.
147. NAC RSC 11306 is the personnel dossier of Mr. Nguyen Van Ngoc, who eventually opened a private medical practice in Phnom Penh in 1938 (see NAC RSC 2116).
148. NAC RSC 6415. Letter to GGI, January 15, 1937.
149. Ibid.
150. NAC RSC 7638.
151. NAC RSC 7638. Emphasis in original.
152. NAC RSC 26695.
153. See Klaus, *Every Child a Lion: The Origins of Maternal and Infant Health Policy in the United States and France, 1890–1920*, chapter 5.
154. NAC RSC 42.
155. NAC RSC 2118, 7637.
156. NAC RSC 7637.
157. NAC RSC 2116. Kompong Cham annual report 1938. Indeed, not only did these French-trained midwives *not* inspire confidence, they may have evoked suspicion. One brief note in the files indicates that several decades earlier, in 1912, a French-trained midwife was killed by a crowd because it was believed that she had been casting evil spells. NAC RSC 2414.
158. Etienne Aymonier, in his personal notes, observed that native midwives generally had elevated reputations in their respective communities. Archives of the Société Asiatique. Fonds Aymonier.
159. NAC RSC 2118.

160. Ibid.
161. Ibid.
162. Declerq et al., "Where to Give Birth," p. 72.
163. See White, "Crossing the River." Cambodian birthing traditions similar to those described in the 1920s were even performed by diasporic communities in the United States in the 1980s. See Carolyn Sargent and John Marcucci, "Aspects of Khmer Medicine among Refugees in Urban America," *Medical Anthropology Quarterly*, new series 16, no. 1 (1984).
164. See also Charlotte Borst, *Catching Babies* (Cambridge, MA: Harvard University Press, 1995).
165. Oakley, *Captured Womb*, p. 2.
166. Norma Sullivan, *Masters and Managers: A Study of Gender Relations in Urban Java*, Women in Asia Publication Series (St. Leonards, Australia: Allen & Unwin, 1994).
167. Ibid., p. 9.
168. Although Southeast Asian studies may identify the unifying (uniform) nature of high female status throughout the region, there are no detailed comparisons between ethnic groups in specific regional studies.
169. Anthony Reid, *Southeast Asia in the Age of Commerce: 1450–1680, Volume One: The Lands Below the Winds*, vol. 1 (New Haven: Yale University Press, 1988), p. 146.
170. Andaya, "Localising the Universal: Women, Motherhood and the Appeal of Early Theravada Buddhism," p. 30.
171. Koven and Michel, "Womanly Duties," p. 1084.
172. Klaus, *Every Child a Lion: The Origins of Maternal and Infant Health Policy in the United States and France, 1890–1920*, chapter 5.
173. Bridie Andrews has written of Chinese revolutionary Qui Jin's efforts to encourage Chinese women into Western-style nursing in this same period: "Nursing as a profession was not an aim in itself, but was seen as a potential means to achieve economic emancipation for women and to contribute to the creation of a modern state in China." Bridie Andrews, "From Bedpan to Revolution: Qui Jin and Western Nursing," in *Women and Modern Medicine*, ed. Lawrence Conrad and Anne Hardy (Amsterdam: Editions Rodopi B. V., 2001), p. 53.

CHAPTER SIX

1. For a discussion of degeneration and depopulation, see chapter 3.
2. Today, medical practitioners refer to leprosy as Hansen's disease, after Armauer Hansen, the researcher who discovered the bacteriological agent. This renaming is not unique to leprosy. A host of diseases underwent redefinition in this period: for example, phthisis and consumption became tuberculosis and croup became diphtheria.
3. Hansen discovered the *Mycobacterium leprae* bacillus in 1873, nearly a decade before Koch discovered the tuberculosis bacillus. The bacterium to this day cannot be cultured in vitro. In the clinic, it has been successfully cultured in vivo only on the footpads of nude mice and on nine-banded armadillos.
4. Leprosy usually first presents with skin lesions that have decreased sensation, numbness in the extremities, and muscle weakness. Tuberculoid leprosy can cause tissue swelling and redness. The more severe lepromatous leprosy can cause skin nodules, lesions, and other deformities from destruction of tissues with mucus membranes. Lepers are highly susceptible to loss of extremities due in part to injury from nerve

damage and gangrene. Because of the bacterium's remarkably slow division rate, symptoms take years to develop.

5. For a more detailed discussion of narrative, emplotment, and leprosy, see Sokhieng Au, "The King with Hansen's Disease: Tales of the Leper in Colonial Cambodia," in *Songs at the Edge of the Forest: Essays in Honor of David Chandler*, ed. Ann Ruth Hansen and Judith Ledgerwood (Ithaca, NY: Cornell University Press, 2008).
6. For general historical reviews, see Michel Foucault, *Madness and Civilization: A History of Insanity in the Age of Reason*, trans. Richard Howard (New York: Vintage Books ed., 1988, c1965); also Zachary Gussow, *Leprosy, Racism, and Public Health: Social Policy in Chronic Disease Control* (Boulder: Westview Press, 1989), particularly the introduction and chapter 1.
7. A leprosarium (*leproserie*) was akin to a *tuberculoserie*. It was used to denote an institution that contained or quarantined lepers, usually a leper village or a leper hospital.
8. NAC RSC 1898, 2174.
9. The links between leprosy, martyrdom, and the church are common in Western European literature. Mary Douglas, "Sacred Contagion," in *Reading Leviticus: A Conversation with Mary Douglas* (Sheffield, UK: Sheffield Academic Press, 1996); Rita Smith Kipp, "The Evangelical Uses of Leprosy," *Social Science & Medicine* 39, no. 2 (1994). In the immediate period of study, we have the widely publicized and sensationalized death of Father Damien in the Molokai, Hawaii leper colony in 1889. For a comparable religious figure in Indochina, see Louis Raillon and Madeline Raillon, *Jean Cassaigne, la lèpre et Dieu* (Paris: Editions Saint-Paul, 1993).
10. See Gussow, *Leprosy, Racism, and Public Health*, introduction.
11. Lucien Gaide and Henry Bodet, "La prévention et le traitement de la lèpre en Indochine," in *Exposition Coloniale Internationale*, ed. Indochine Française. Section des Services d'Intérêt Social. Inspection Générale des Services Sanitaires et Médicaux de l'Indochine (Hanoi: Imprimerie d'Extrême Orient, 1931).
12. Jean Vaudon, "Les Filles de Saint-Paul de Chartes: Une léproserie en Indochine," *Revue d'histoire des missions* 3, no. 4 (1926). The author claims that the leprosarium was actually controlled by the Sisters of Saint Paul when it opened. The colonial documentation is mute on this claim.
13. For a longer version of this story, see Au, "The King with Hansen's Disease." Multiple versions of the story have been recorded by French and Khmer scholars. Angier, "La lèpre au Cambodge: Croyances et traditions," *Annales d'hygiène et de médecine coloniales* 6 (1904), p. 176–77; David Chandler, "Folk Memories of the Decline of Angkor in Nineteenth-Century Cambodia: The Legend of the Leper King," *Journal of the Siam Society* 61, no. 1 (1979), p. 55; A. Kermorgant, "Notes sur la lèpre dans nos diverses possessions coloniales," *Annales d'hygiène et de médecine coloniales* 8 (1905): p. 634, Ritharasi Norodom, "L'évolution de la médecine au Cambodge" (M.D. thesis, Librarie Louis Arnett, 1929), pp. 66–68. Etienne Aymonier, in his personal notes of 1884, seems to have recorded one of the earliest versions of the leper king folk story. Archives of the Societé Asiatique. Aymonier collection. (The Aymonier archives have yet to be systematically catalogued.)
14. Chandler, "Folk Memories of the Decline of Angkor," passim.
15. Mark Gregory Pegg, "Le corps et l'autorité: La lèpre de Baudoin IV," *Annales: Economies sociétés civilisations* 45, no. 2 (1990). Pegg argues that in Western Christendom, a leper could never be crowned king because of the association with moral unworth. In contrast, the Latin Kingdom of Jerusalem in 1174 anointed and crowned the leper Baldwin IV.

16. See Au, "The King with Hansen's Disease."
17. Solange Thierry, *Le Cambodge des contes* (Paris: Editions L'Harmattan, 1985), chapter 16. According to Thierry's analysis of Khmer oral tales, the three levels of power—magical, meritorious, and royal—constantly interact and direct the fate of characters in these stories.
18. Angier, in his 1897 study, would observe that the story of the leper king was well-known throughout the countryside. NAC RSC 2174.
19. Tcheou Ta-Kouan, *Mémoires sur les coutumes du Cambodge,* trans. Paul Pelliot (Paris: Librarie d'Amérique et d'Orient, 1951), p. 23.
20. Dr Angier, "Le Cambodge: Géographie médicale," *Annales d'hygiène et de médecine coloniales* 4 (1901): p. 33.
21. See, for example, Angier, "La lèpre au Cambodge"; Charles Meyer, "Lepers in Troeung," *Études cambodgiennes* 13 (1968). See also NAC RSC 144, 1898, 2174, 12623.
22. NAC RSC 2174.
23. In a recent study on Thailand, researcher Liora Navon found from informant interviews that before the 1950s, sufferers of leprosy encountered ambivalent rather than severely stigmatizing reactions. Liora Navon, "Beggars, Metaphors, and Stigma: A Missing Link in the Social History of Leprosy," *Social History of Medicine* 11, no. 1 (1998).
24. The statue of the leper king at Angkor Thom is a reproduction. The original is at the National Museum in Phnom Penh.
25. Pierre Benoit, *Le roi lépreux* (Paris: Albin Michel, 1927), p. 117. Also peculiar is his ascription of "height slightly below average" to a solitary figure in a seated position.
26. Ibid., p. 118.
27. This is not only a French interpretation. British writer Perry Burgess would write of the figure in 1936, "I saw the statue of the old boy sitting disconsolately and alone . . . , his only companions the monkeys and birds by day and the prowling panther at night." Perry Burgess, "Lepers and Leprosy," *Scientific Monthly* 42, no. 5 (1936), p. 399.
28. NAC RSC 739.
29. NAC RSC1375.
30. See Peter Zinoman, *The Colonial Bastille: A History of Imprisonment in Vietnam, 1862–1940* (Berkeley: University of California Press, 2001), pp. 94–97. To my knowledge, no detailed study has been done of prisoner health in Cambodia or Indochina generally; however, a cursory review of the RSC medical files relating to prisons reveals an appalling death rate, with frequent epidemics and high morbidity.
31. No clear evidence existed for the comparative levels of contagion of these two diseases in the first decade of the twentieth century. However, in some of the medical literature there was tacit understanding that tuberculosis appeared to be more contagious.
32. For an explanation of the heightened concern over leprosy at this time, see Gussow, *Leprosy, Racism, and Public Health,* particularly p. 114; Sanjiv Kakar, "Leprosy in British India, 1860–1940: Colonial Politics and Missionary Medicine," *Medical History* 40 (1996); Eric Silla, *People Are Not the Same: Leprosy and Social Identity in Twentieth-Century Mali* (Portsmouth, NH: Heinemann, 1998), introduction.
33. NAC RSC 2174. Population numbers are taken from Jean Delvert's estimates. Jean Delvert, *Le paysan cambodgien* (Paris: Mouton, 1961).
34. Edouard Jeanselme, *Étude sur la lèpre dans la péninsule indo-chinoise et dans le Yunnan* (Paris: Carré et Naud, 1900), p. 76.

35. A. Kermorgant, "Maladies epidémiques et contagieuses qui ont régné dans les colonies françaises en 1901," *Annales d'hygiène et de médecine coloniales* 6 (1903): p. 606.
36. Kermorgant, "Notes sur la lèpre," p. 33.
37. NAC RSC 12623.
38. NAC RSC 12623.
39. NAC RSC 12623.
40. See chapter 5.
41. NAC RSC 12623. A 1910 report from the inspector general notes that leprosy was unequally distributed. Most cases were on the Red River delta in Tonkin, on the Mekong delta of Cochinchina, and at the confluence of the two rivers in Annam. He notes that the number of lepers in Cambodia and Laos was insignificant. See also NAC RSC 111.
42. See NAC RSC 118, 12505.
43. In the French medical literature of the turn of the century, the concept of *hérédocontagion* served as a sort of conceptual meeting ground between the hereditarians and the contagionists for diseases such as leprosy, syphilis, and tuberculosis.
44. Primet, "Rapport sur la prophylaxie de la lèpre en Indochine," (Hanoi-Haiphong: Gouvernement Général de L'Indochina, Direction Générale de la Santé, 1910), p. 2.
45. J. C. Boscq, "Assistance Médicale en Indochine: La leproserie de Cu-Lao-Rong," in *La dépêche coloniale illustree: La vie econonomique de nos colonies*, ed. Colonial Rédaction et Administration (Paris: Rédaction et Administration, 1909), p. 131. The first year of this leprosarium's operation seemed rather inauspicious. According to a Catholic journal, six months after Culao Rong opened, it was hit by a cyclone. Two lepers were killed by falling debris, and an outbreak of cholera and gangrene soon followed. Vaudon, "Les filles de Saint-Paul," p. 545. Without a local leprosarium, the Cambodian government sent a handful of lepers to Culao-Rong annually until 1915. In that year, the Cambodian leprosarium of Troeung opened. NAC RSC 743 and 2025.
46. NAC RSC 25608.
47. See chapter 4 for a discussion of the control of epidemics in Cambodia.
48. For a discussion of *dépistage*, see chapter 4.
49. NAC RSC 2174.
50. David S. Barnes, *The Making of a Social Disease: Tuberculosis in Nineteenth-Century France* (Berkeley: University of California Press, 1995), introduction.
51. Edouard Boinet, "La lèpre à Hanoi (Tonkin)," *Revue de médecine* 10 (1890), p. 611. This is highly improbable, as infection usually requires prolonged close contact with bacteriologically-active infected individuals. Also, it usually takes four or more years after exposure for the first symptoms to appear.
52. Boscq, "Assistance Médicale en Indochine," p. 135.
53. NAC RSC 2174. Angier had observed that leprosy tallies were confounded for several reasons. The indigenous population did not consider the early stages of leprosy a disease; however, this underreporting was offset by the tendency to lump other chronic skin conditions with the affliction.
54. Foucault argues that the concept of a disease "essence" disappeared in France with the birth of the clinic in the late eighteenth century, and was replaced by a geographical and normative concept of disease, a "complex movement of tissues in reaction to an irritating cause." This is "site" replacing "symptomology." He writes of the transformation of illness and disease, "It is no longer the pathological species inserting itself into the body wherever possible; it is the body itself that has become ill."

Michel Foucault, *The Birth of the Clinic* (New York: Random House, 1994), p. 136. I would make a hypothesis here that, if these processes of change were indeed profound within the medical field, germ theory could be seen as a *bouleversement* (shift back) or perhaps a reformed sort of disease essentialism.

55. Confirmed cases were considered cases testing bacterially positive in the laboratory.
56. NAC RSC 1513.
57. NAC RSC 2118.
58. NAC RSC 10159.
59. As a comparison, see CAOM INDO RSTNF file 3683 for evasion statistics for Tonkin leprosaria.
60. Boscq, "Assistance Médicale en Indochine," p. 134.
61. For more information on the SPE and the métis population, see chapter 5.
62. The Catholic village was on the north side of Phnom Penh. For more information, see chapter 4.
63. NAC RSC 1513.
64. NAC RSC 1513. Medical certificate issued April 26, 1934.
65. NAC RSC 1513. Letter from the governor of Cochinchina to the RSC, May 21, 1934.
66. Ann Laura Stoler, "Sexual Affronts and Racial Frontiers," in *Tensions of Empire: Colonial Cultures in a Bourgeois World*, ed. Ann Laura Stoler and Frederick Cooper (Berkeley: University of California Press, 1997). See also J. Mazet, *La condition juridique des métis dans les possessions françaises* (Paris: Editions Domat-Montchrestien, 1932).
67. We have only one record of a full European who developed leprosy in Cambodia; he was quickly repatriated to France. NAC RSC 1337.
68. Two years later, however, another SPE student, Joseph Boulet, was interred at Troeung with little discussion of his racial status (if considerable discussion over financial responsibility). The SPE agreed to pay for his internment with that of another young boy named Chha. A student of the SPE, Chha is a puzzling case because both of his parents were known and had Khmer names. Boulet's internment with Chha would suggest that Boulet may have been legally classified as Indochinese. NAC RSC 1497. It is possible that Chha was *métis* in appearance, but recognized by his mother's Khmer husband as his child.
69. NAC RSC 10160.
70. See the annual reports for Kompong Cham, particularly NAC RSC 87, 92, 630, 1729, 2486, 12635, and 26695.
71. Boscq, "Assistance Médicale en Indochine"; Gaide and Bodet, "La prévention," p. 11.
72. NAC RSC 1356.
73. Boinet, "La lèpre à Hanoi," pp. 618–19.
74. Thiroux, "Contribution à l'étude de la contagion et de la pathogénie de la lèpre," *Annales d'hygiène et de médecine coloniales* 6 (1903): p. 564. The story is most likely spurious. Such a fast rate of development for Hansen's disease is rare. Current estimates are that leprosy takes approximately four to twenty years from the time of first exposure to present its initial symptoms.
75. Ibid.: p. 573.
76. Chaulmoogra is an oil obtained from the seed of the tree *Hydrocarpnus kurzii*, commonly found in India.
77. Chapter 2 discusses the state of international biomedical research.
78. In parts of French colonial Africa, the AM adopted injections of gorli nut oil as a standard leprosy treatment; this oil was also a traditional leprosy medicine used in Guinea and the Ivory Coast. Silla, *People Are Not the Same*, chapter 1.

79. NAC RSC 1337. Letter to RSC, February 12, 1922, on successful (Vietnamese) treatment, with a preparation recipe and a recommendation from the governor of Cochinchina. CAOM INDO RSTNF. In Te Truong, Tonkin, a treatment from Bombay was tested on internees.
80. For example, Dr. Baillon of Prey Veng openly admitted in 1922 that he regularly consulted a Kru Khmer when treating lepers. NAC RSC 1814.
81. NAC RSC 2174, 2465. In one example, a young girl died from secondary infections after burns obtained when she underwent an indigenous treatment for leprosy at a *wat* in 1921. NAC RSC 8841.
82. NAC RSC 23150.
83. The Cambodian krabao tree is known by the scientific name *Hydnocarpus anthelminthicus*. It is of the same genus as the chaulmoogra tree.
84. An actual recipe for this krabao treatment was recorded by Dr. Menaut. Bernard Menaut, "Matière médicale cambodgienne," in *Exposition Coloniale Internationale*, ed. Inspection Générale des Services Sanitaires et Médicaux de l'Indochine. (Hanoi: Imprimerie d'Extrême-Orient, 1931), p. 23.
85. NAC RSC 49, 84, 112, 115, 1349, 11839.
86. NAC RSC 630, 7808
87. NAC RSC 2115, 2465.
88. NAC RSC 630.
89. NAC newspaper collections. *La Dépêche*, "La lutte contre la lepre," 4/27/1935. For more information on the development of synthetic drugs for leprosy, see John Parascandola, "Miracle at Carville: The Introduction of the Sulfones for the Treatment of Leprosy," *Pharmacy in History* 40, no. 2 (1998).
90. This oxcart trail would, by 1930, become the colonial road between Kompong Cham and Phnom Penh.
91. Menaut, "Matière médicale cambodgienne," pp. 21–23. See also NAC RSC 4946, 10160.
92. Several years later, Menaut would inconsistently note that there were 107 lepers when he found the village. NAC RSC 4946.
93. Menaut, "Matière médicale cambodgienne," pp. 21–23.
94. NAC RSC 10159, 10160.
95. NAC RSC 10159.
96. NAC RSC 10159.
97. ". . . fut sa ruin." Menaut, "Matière médicale cambodgienne," p. 22.
98. The restriction against spending the night, although seemingly mild, would have severely hampered visitation of any sort. Travel from Kampong Cham, only fifteen kilometers away, could take two or more days before the colonial roads were built. The Khmer taboo against spending the night in the "wilderness" would make a day trip unworkable.
99. Births were rare at Troeung Leprosarium, probably due to the low ratio of women to men. I have been unable to locate any records on confiscated babies in Cambodia. However, in neighboring Tonkin, 569 infants were taken away from their leprous parents between 1912 and 1929. In a 1931 report, the colonial doctor would observe that to that year, 72.2 percent of these confiscated children had died of "intercurrent maladies." Despite this shockingly low survivorship, the doctor would note in self-congratulation that none of the remaining children showed any signs of leprosy. Gaide and Bodet, "La prévention," p. 43.
100. NAC RSC 10159. June 7, 1915, decree.

101. NAC RSC 10159. Nov. 10, 1916, decree. These meals were to be served at 7:00 a.m., 11:00 a.m., and 6:00 p.m. Each adult was to be provided every day with .7 kg of second-grade white rice, .2 kg of fish or .15 kg of other meat, .15 kg of vegetables, .02 kg of nuoc-mam, and .003 kg of tea. The quantity of rations was reduced proportionally by age.
102. NAC RSC 10159.
103. NAC RSC 2481, 10159, 10160.
104. NAC RSC 10159. Milder regulatory infractions could lead to a suppression of a portion of daily rations.
105. NAC RSC 10160. Koh Norung, a suitable island in proximity to Kompong Cham town, was allocated to contain this leprosarium. In Tonkin, more drastic although equally ineffective barriers, such as trenches and thick hedges around leprosaria, were implemented. CAOM BIB 5006.
106. NAC RSC 2465.
107. See Susan L. Burns, "From 'Leper Villages' to Leprosaria: Public Health, Nationalism and the Culture of Exclusion in Japan," in *Isolation: Places and Practices of Exclusion*, ed. Alison Bashford and Carolyn Strange (New York: Routledge, 2003); Kakar, "Leprosy in British India, 1860–1940"; Diana Obregón, "Building National Medicine: Leprosy and Power in Colombia, 1870–1910," *Social History of Medicine* 15, no. 1 (2002). Other colonies of Southeast Asia were also creating government-sponsored leper villages at the turn of the century. Warwick Anderson, "Leprosy and Citizenship," *Positions* 6, no. 3 (1998); Executive Committee of the Eighth Congress of the Far Eastern Association of Tropical Medicine, *Siam in 1930: General and Medical Features*, 2nd ed. (Bangkok: White Lotus Press, 2000 [1930]), chapter 23; Loh Kah Seng, "'Our Lives Are Bad but Our Luck Is Good': A Social History of Leprosy in Singapore," *Social History of Medicine* 21, no. 2 (2008). See also, for a comparison of different internment policies in the United States and Norway, Gussow, *Leprosy, Racism, and Public Health*.
108. Anderson, "Leprosy and Citizenship," p. 721.
109. Ibid., p. 718.
110. The European preoccupation with such disciplinary transformation in this same period is exactly what Foucault originally studied in this regard. However, his analysis assumed that the discourse was embedded within the society it normalized.
111. NAC RSC 21239. Ittiacandy's note.
112. The *balaat* is the indigenous lieutenant governor of a province.
113. Ek's declaration, dated June 26, 1917, in Khmer. NAC RSC 21239.
114. An administrative province is a *khaet*. It is subdivided into *srok*, then *khum* (large village or groups of hamlets), then *phum* (hamlet). The *srok* and *khum* have administrators termed *mesrok* and *mekhum*. See the abbreviations and glossary sheet.
115. NAC RSC 21239.
116. Menaut, "Matière médicale cambodgienne," p. 23.
117. NAC RSC 10160. This opinion was seconded the following year with observations of general filth and misery in the village. NAC RSC 2481.
118. NAC RSC 1337, 1497, 1513.
119. See NAC RSC annual medical reports for Kompong Cham, particularly 87, 92, 630, 1729, 2486, 12635, and 26695.
120. NAC RSC 2465.
121. NAC RSC 2465 and 26695.
122. NAC RSC 26695.

123. NAC RSC 2465.
124. NAC RSC 23150.
125. For example, in the relatively large government leprosarium of Te Truong (five hundred inhabitants) in neighboring Vietnam, the villagers were permitted to run an internal market five days a week. Although this market was exclusively for lepers, the administration still enforced strange rules to disinfect the money. The money boxes of merchants were specially built, "divided in two compartments with a [windowed] partition." One compartment held the coin, while the other held several tablets of trioxymethylene. The open window ideally exposed all moneys passing through the market to these vapors for several hours. This again reflects the extreme paranoia that leprosy induced. CAOM INDO RSTNF 3745. The director of the leprosarium would note that some objects, particularly chicken and duck eggs, seemed to be the "object of a clandestine commerce."
126. NAC RSC 23150. Letter from the *résident* of Kompong Cham to the RSC on March 7, 1934.
127. NAC RSC 26695. Kompong Cham annual report for 1934, p. 11.
128. NAC RSC 10159. Letter to the RSC, November 19, 1914.
129. NAC RSC 12623.
130. NAC RSC 630.
131. NAC RSC 129, 133, 630, 4950, 12505, 12623.
132. NAC RSC 1513.
133. NAC RSC 21115. By 1942, leper internments were billed to religious congregations, families, and other small groups. NAC RSC 833. By this point, the Vichy regime had been installed in France and colonial medical record keeping had drastically declined, so we cannot determine whether this affected official leper declarations or community declarations.
134. The RSC was not alone in its failure at operating a leprosarium. Vietnam witnessed several leper uprisings and complaints in its official leprosaria during the 1930s and 1940s. For instance, the leprosarium of Phu-Tho had a coordinated mass escape of ten individuals when the lepers received no official relief after complaining of exploitation, forced labor, extortion, and monthly allowances stolen by the village chief and Catholic Father Tuan. CAOM NTF 3745.
135. NAC RSC 605 and 2117 discuss the failure of a contractor to finish construction of new buildings at Troeung in 1938 and 1939 after taking payment.
136. Menaut, "Matière médicale cambodgienne," p. 23.
137. NAC RSC 144, 1513, 26696.
138. B. Menaut and H. Baisez, *La lèpre au Cambodge* (Hanoi: Imprimerie d'Extrême-Orient, 1919), p. 95. See also NAC RSC 1356.
139. NAC RSC 118.
140. For a history of the development of these treatments, see Parascandola, "Miracle at Carville: The Introduction of the Sulfones for the Treatment of Leprosy."
141. Meyer, "Lepers in Troeung," passim.

CHAPTER SEVEN

1. See Eric Thomas Jennings, *Vichy in the Tropics: Pétain's National Revolution in Madagascar, Guadeloupe, and Indochina, 1940–1944* (Stanford, CA: Stanford University Press, 2001), chapters 6 and 7; Martin Thomas, *The French Empire at War, 1940–45* (Manchester, UK: Manchester University Press, 1998). French governance of Cam-

bodia between 1945 and 1954 was much more limited than its control from 1882 to 1939.

2. For a short review, see A. J. Stockwell, "Southeast Asia in War and Peace: The End of European Colonial Empires," in *The Cambridge History of Southeast Asia: Volume 2 Part Two: From World War II to the Present*, ed. Nicholas Tarling (Cambridge: Cambridge University Press, 1992).
3. Jean-Pierre Dedet, *Les Instituts Pasteur d'outre-mer: Cent vingt ans de microbiologie française dans le monde* (Paris: Harmattan, 2000).
4. Consider the strange article by Tan at the late date of 1981, discussing the quinquina development program that had "stalled" in 1966, and the need to restart it. Boun Suy Tan, "Introduction du quinquina au Cambodge," *Seksa Khmer* 1, no. 3 (1981). In 1966, quinine was already outmoded as a treatment for malaria.
5. Cambodian Government, *Journal officiel du Cambodge* (Phnom Penh: Cambodian Government, 1955).
6. See Clifford Geertz, *Negara: The Theatre State in Nineteenth-Century Bali* (Princeton, NJ: Princeton University Press, 1980).
7. Gyan Prakash, *Another Reason: Science and the Imagination of Modern India* (Princeton, NJ: Princeton University Press, 1999), p. 34.
8. Michael J. Heffernan, "A State Scholarship: The Political Geography of French International Science during the Nineteenth Century," *Transactions of the Institute of British Geographers* 19, no. 1 (1994).
9. See Philip D. Curtin, *Death by Migration: Europe's Encounter with the Tropical World in the Nineteenth Century* (Cambridge, MA: Cambridge University Press, 1995). For a discussion of the relationship between the environment more generally and European expansion, see David Arnold, *The Problem of Nature: Environment, Culture, and European Expansion*, ed. Constantin Fasolt, New Perspectives on the Past (Oxford: Blackwell Publishers, 1996).
10. All categories are false in that we impose them using arbitrary sets of characteristics. The predominance in "the tropics" of many tropical diseases thus seems a reasonable way to classify them. However, there are some strange inclusions—for instance leprosy, classified as a tropical disease, is common to many temperate and sub-Arctic regions of the world. See Alison Bashford, *Imperial Hygiene: A Critical History of Colonialism, Nationalism and Public Health* (New York: Palgrave Macmillan, 2004), chapter 4.
11. Scholars rightly tie tropical medicine to colonial formulations of knowledge—but, in my opinion, they wrongly claim that these links make the separate category of tropical medicine false. See the collected essays in David Arnold, ed., *Warm Climates and Western Medicine: The Emergence of Tropical Medicine, 1500–1900* (Atlanta: Rodopi, 1996).
12. As discussed in the introduction, the process of statistical collection was itself also inconsistent.
13. Thomas McKeown, *The Modern Rise of Population* (New York: Academic Press, 1976).
14. See chapter 4.
15. Anne Hardy, in a study of nineteenth-century Britain, examines each of these theses in relation to the epidemiological histories of specific diseases. She argues that nutrition is not the key factor in reducing mortality, but rather that factors such as stability, altered behaviors, and certain public-health measures determine population

growth. Anne Hardy, *The Epidemic Streets: Infectious Disease and the Rise of Preventive Medicine, 1856–1900* (Oxford: Clarendon Press, 1993).

16. Consider that, in the Philippines, American colonialism heralded a devastating spike in disease, epidemic, and death. Yet the Philippines still experienced population growth throughout the turn of the century. Ken De Bevoise, *Agents of Apocalypse: Epidemic Disease in the Colonial Philippines* (Princeton, NJ: Princeton University Press, 1995).
17. Anthony Reid, *Southeast Asia in the Age of Commerce: 1450–1680, Volume One: The Lands Below the Winds*, 2 vols., vol. 1 (New Haven: Yale University Press, 1988), p. 18.
18. This is stability not in cultural forms or traditional life, but in terms of protections from massive displacement and death caused by open civil war.
19. John Sydenham Furnivall, *Netherlands India* (Cambridge: Cambridge University Press, 1944), p. 446. Carl Trocki expands on this formulation, observing that these plural societies also have sexual and psychological dimensions. Polyglot communities with different social niches that preexisted colonial intervention became confounded with political and social roles in the colonial era. Nicholas Tarling, ed., *The Cambridge History of Southeast Asia: Volume 2 Part One: From c. 1800 to the 1930s*, 2 vols., vol. 2 (Cambridge: Cambridge University Press, 1992), chapter 2, p. 109.
20. Mandala is a term for the political kingdoms of middle-period Southeast Asia: waxing and waning strong political centers with peripheries giving tribute. These kingdoms were highly cyclical, territorially overlapping, and based on various types of influence. Reviews are provided in the research of Charles Higham and O. W. Wolters. Charles Higham, *The Archaeology of Mainland Southeast Asia* (Cambridge: Cambridge University Press, 1989); O. W. Wolters, *History, Culture and Region in the Southeast Asian Perspectives* (Pasir Panjang, Singapore: Institute of Southeast Asian Studies, 1982).
21. Charles E. Rosenberg, *The Care of Strangers: The Rise of America's Hospital System* (New York: Basic Books, Inc., 1987), p. 7; Paul Starr, *The Social Transformation of American Medicine* (New York: Basic Books, Inc., 1982). One must realize, particularly with the increasing popularity of holistic and alternative medicine in the last few decades in America, that the doctor's authority is also not absolute in the West. It is a matter of degree.
22. Paul U. Unschuld, *Medicine in China: A History of Ideas* (Berkeley: University of California Press, 1985), p. 249.
23. Anne Guillou's dissertation, particularly part 3, deals with "how Khmer patients negotiate the perceived foreignness of medicine offered by modern Cambodian doctors." Anne Yvonne Guillou, "Les médecins au Cambodge: Entre élite sociale traditionnelle et groupe professionel moderne sous influence étrangère" (PhD thesis, Ecole des Hautes Etudes en Sciences Sociales, 2001).
24. Ibid., pp. 314–15. These statistics are for Kandal.
25. Immediately after independence, Vietnam launched a medical program promoting a Vietnamese brand of scientism. It was to be built on "scientific, national, and popular principles." In a speech to the Vietnamese health corps in February 1955, Ho Chi Minh declared, "Our ancestors possessed much useful knowledge about the treatment of illness . . . . To increase the sphere of action of medicine, it is indispensable for us to study ways in which to ally Oriental treatments with those of the Occident." Bao Chau Hoang, "L'alliance des deux médecines traditionnelle et moderne au Vietnam," *Etudes vietnamiennes*, special issue on *La Médecine Traditionelle*, no. 50 (1977):

p. 7. In other countries of postcolonial Southeast Asia, many leaders made strong ideological commitments to a "local" version of science.

26. In recent years the Cambodian government has discussed the need for drug regulation, but one can still walk into a market and buy any range of powerful drugs without authorization.
27. See, for example, C. H. Bledsoe and M. F. Goubaud, "The Reinterpretation of Western Pharmaceuticals among the Mende of Sierra Leone," *Social Science and Medicine* 21, no. 3 (1985). This is of one of the earliest of many anthropological studies of this phenomenon.
28. Sjaak van der Geest, Susan Reynolds Whyte, and Anita Hardon, "The Anthropology of Pharmaceuticals: A Biographical Approach," *Annual Review of Anthropology* 25 (1996): pp. 166–70.
29. For a key anthropological text, see Nicholas Thomas, *Entangled Objects: Exchange, Material Culture, and Colonialism in the Pacific* (Cambridge, MA: Harvard University Press, 1991). For a recent science-studies volume, see Lorraine Daston, *Things That Talk: Object Lessons from Art and Science* (New York: Zone Books, 2004).
30. See, for example, Byron J. Good, *Medicine, Rationality, and Experience: An Anthropological Perspective*, ed. Anthony T. Carter, The Lewis Henry Morgan Lectures: 1990 (Cambridge: Cambridge University Press, 1994); Nancy Scheper-Hughes and Margaret Lock, "Mindful Body: A Prolegomenon to Future Work in Medical Anthropology," *Medical Anthropology Quarterly*, new series 1, no. 1 (1987); Brian S. Turner, *Medical Power and Social Knowledge*, 2nd ed. (London: Sage Publications, 1995).
31. Along with "system-wide" studies of medical practices such as Rosenberg's *The Care of Strangers* and J. H. Warner's study of nineteenth-century America, we have excellent histories of more limited social phenomena that have strongly affected medical practices: for instance, the collected essays in Ann La Berge and Mordechai Feingold, eds., *French Medical Culture in the Nineteenth Century* (Atlanta: Editions Rodopi, 1994). See also Nancy Tomes, *The Gospel of Germs: Men, Women and the Microbe in American Life* (Cambridge, MA: Harvard University Press, 1998); John Harley Warner, *The Therapeutic Perspective: Medical Practice, Knowledge and Identity in America, 1820–1885* (Princeton, NJ: Princeton University Press, 1997).
32. Anthony Fauci, "Smallpox Vaccination Policy: the Need for Dialogue," *New England Journal of Medicine* 346, no. 17 (2002), p. 1319.
33. NAC RSC 1315, 1346.
34. Standard humanitarian health relief manuals such as the SPHERE handbook repeatedly urge the inclusion of these local community leaders (formerly known as elites) in the establishment of "sustainable" programs.
35. For an amusing account of what even today continues to be the model of participation, see B. D. Paul and W. J. Demarest, "Citizen Participation Overplanned: The Case of a Health Project in the Guatemalan Community of San Pedro La Laguna," *Social Science and Medicine* 19, no. 3 (1984).
36. NAC RSC 1248.
37. W. Sorensen et al., "Assessment of Condom Use among Bolivian Truck Drivers through the Lens of Social Cognitive Theory," *Health Promotion International* 22, no. 1 (2007), p. 41.

# GLOSSARY

***achaar.*** Lay ritual officiant of the Buddhist temple; not a monk.

**Annamite.** Ethnic Vietnamese.

***balaat.*** Indigenous lieutenant governor of a province.

***chhlong.*** To cross; also used to denote "crossing" of disease from one person to another, and thus as a translation of the French *contagion* into Khmer.

***chhlong tonle.*** Literally, "to cross the lake"; to give birth.

***chhmap.*** Khmer traditional midwife.

***chomngu.*** Illness or disease.

***chomngu khlong.*** Cluster of symptoms usually translated as leprosy.

***infirmier/infirmière.*** Could be translated as "nurse," but these individuals also frequently served as orderlies, pharmacists, guards, translators, and general assistants for the doctors. Most were male. Due to English associations with the word "nurse," I retain the French term.

***kchaal.*** Wind, seen as one of the vital essences of the body that was prone to blockage and could thus cause sickness.

***khaet.*** Province.

***khum.*** Administrative subdivision of *srok*, made up of several *phums*.

***kronh.*** Fever.

***kronh chanh.*** Cluster of symptoms usually translated as malaria.

***kru khmer.*** Literally, Khmer "teacher"; but often used to refer to a Khmer traditional doctor.

***kru peet.*** Hospital-based or Western-trained medical doctor.

***lèpre.*** Leper.

***me.*** Head, master, or chief.

***médecin.*** A medical caregiver of considerably lower rank than a *docteur en médecine*. There is no English equivalent to denote the different degrees, so when the two are used together, I retain the French terms.

***mekhum.*** Chief or head of the *khum.*

***mesrok.*** Head of the *srok.*

***métis.*** Mixed French-indigenous children.

***munti.*** Government building.

***neang.*** Title for a young woman, similar to "Miss."

***neak.*** Title of address for either sex; also means "person." For a woman associated with the royal court, I translate it as "Lady."

***neak-ta.*** Ancestor spirit of a locale; grandfather spirit.

***neak-yay.*** Ancestor spirit of a locale; grandmother spirit.

***peet.*** Hospital or clinic.

***peh.*** Khmer transliteration of French *peste* (plague).

***peste.*** Plague.

***phum.*** Village or hamlet.

***prey.*** Wilderness.

***ruup.*** Spirit medium.

***sangha.*** Buddhist monastic order.

***sasae.*** Threads or strings; in Khmer physiology, denotes pathways or threads in the body through which *kchaal* flows.

***srok.*** Administrative subdivision of *khaet.* More generally, a city, country, or civilized area (in contrast to *prey*).

***vaccinifère.*** Person, usually a child, recently inoculated against smallpox, whose active smallpox sores or pus are used to vaccinate others.

***wat.*** Buddhist temple.

# SOURCES

### ARCHIVAL SOURCES

Archives of the Institut Pasteur, Paris
Archives of the Société Asiatique, Fonds Aymonier, Paris
Bibliothèque Nationale, Paris
Centre des Archives d'Outre Mer, Aix-en-Provence
National Archives of Cambodia, Phnom Penh
National Library of Cambodia, Phnom Penh

### BIBLIOGRAPHY

Académie de Médecine Français. "Rapport générale présenté à M. Le Ministre de l'Intérieur par l'Académie de Médecine sur les vaccinations et revaccinations pratiquées en France et dans les colonies pendant l'année 1899." Melun: Académie de Médecine. Imprimerie Administrative, 1899.

Ackerknecht, Erwin H. "Anticontagionism between 1821 and 1867." *Bulletin of the History of Medicine* 22, no. 5 (1948): 562–93.

———. "Hygiene in France, 1815–1848." *Bulletin of the History of Medicine* 22, no. 2 (1948): 117–55.

———. *Medicine at the Paris Hospital, 1794–1848*. Baltimore: Johns Hopkins Press, 1967.

Ackerman, Evelyn. *Health Care in the Parisian Countryside, 1800–1914*. New Brunswick: Rutgers University Press, 1990.

Adas, Michael. "From Avoidance to Confrontation: Peasant Protest in Precolonial and Colonial Southeast Asia." *Comparative Studies of Society and History* 23, no. 2 (1981): 217–47.

Aisenberg, Andrew R. *Contagion: Disease, Government, and the "Social Question" in Nineteenth-Century France*. Stanford: Stanford University Press, 1999.

Andaya, Barbara. Introduction to *Other Pasts: Women, Gender and History in Early Modern Southeast Asia*, edited by Barbara Andaya, 1–26. Manoa: University of Hawaii Press, 2000.

———. "Localising the Universal: Women, Motherhood and the Appeal of Early Theravada Buddhism." *Journal of Southeast Asian Studies* 33, no. 1 (2002): 1–30.

Anderson, Warwick. "Leprosy and Citizenship." *Positions* 6, no. 3 (1998): 707–30.

Andrew, C. M., and A. S. Kanya-Forstner. "The French Colonial Party: Its Composition, Aims and Influence." *Historical Journal* 14, no. 1 (1971): 99–128.

Andrews, Bridie. "From Bedpan to Revolution: Qui Jin and Western Nursing." In *Women and Modern Medicine*, edited by Lawrence Conrad and Anne Hardy, 53–71. Amsterdam: Editions Rodopi B.V., 2001.

Angier, Henri-Albert. "Le Cambodge: Géographie médicale." *Annales d'hygiène et de médecine coloniales* 4 (1901): 5–59.

———. "La lèpre au Cambodge. Croyances et traditions." *Annales d'hygiène et de médecine coloniales* 6 (1904): 176–80.

Arnold, David, ed. *Warm Climates and Western Medicine: The Emergence of Tropical Medicine, 1500–1900*. Atlanta: Rodopi, 1996.

———. *Colonizing the Body: State Medicine and Epidemic Disease in Nineteenth-Century India*. Berkeley: University of California Press, 1993.

———. *The Problem of Nature: Environment, Culture, and European Expansion*. Edited by Constantin Fasolt. Oxford: Blackwell Publishers, 1996.

Au, Sokhieng. "Indigenous Politics, Public Health and the Cambodian Colonial State." *South East Asia Research* 14, no. 1 (2006): 33–86.

———. "The King with Hansen's Disease: Tales of the Leper in Colonial Cambodia." In *Songs at the Edge of the Forest: Essays in Honor of David Chandler*, edited by Ann Ruth Hansen and Judith Ledgerwood, 121–136. Ithaca, NY: Cornell University Press, 2008.

Aymonier, Etienne. *Notes sur les coutumes et croyances superstitieuses des cambodgiens*. Paris: CEDORECK, 1984 (1883).

Baldwin, Peter. *Contagion and the State in Europe, 1830–1930*. Cambridge: Cambridge University Press, 1999.

Baradat, R. "Le Samrê ou Péâr: Population primitive de l'ouest du Cambodge." *BEFEO* XLI, no. 1 (1941): 1–148.

Bareau, André. "Une representation du monde selon le tradition bouddhique." *Seksa Khmer* 1, no. 5 (1982): 11–16.

Barnes, David S. *The Making of a Social Disease: Tuberculosis in Nineteenth-Century France*. Berkeley: University of California Press, 1995.

Bartholomew, James R. *The Formation of Science in Japan: Building a Research Tradition*. New Haven: Yale University Press, 1989.

Bashford, Alison. *Imperial Hygiene: A Critical History of Colonialism, Nationalism and Public Health*. New York: Palgrave Macmillan, 2004.

Bates, Don, ed. *Knowledge and the Scholarly Medical Traditions*. Cambridge: Cambridge University Press, 1995.

Bayly, Susan. "French Anthropology and the Durkheimians in Colonial Indochina." *Modern Asian Studies* 34, no. 3 (2000): 581–622.

Benoit, Pierre. *Le roi lepréux*. Paris: Albin Michel, 1927.

Berlinguer, Giovanni. "Globalization and Global Health." *International Journal of Health Services* 29, no. 3 (1999): 579–95.

Bernal, Martin. "The Nghe-Tinh Soviet Movement 1930–1931." *Past and Present*, no. 92 (1981): 148–68.

Betts, Raymond F. *Assimilation and Association in French Colonial Theory 1890–1914*. New York and London: Columbia University Press, 1961.

Bitard, Pierre. "Le monde du sorcier au Cambodge." In *Le monde du sorcier*, 307–27. Paris: Editions du Seuil, 1966.

Bizot, François. *Le figuier à cinq branches: Recherche sur le bouddhisme khmer*. Paris: EFEO, 1976.

Bledsoe, C. H., and M. F. Goubaud. "The Reinterpretation of Western Pharmaceuticals among the Mende of Sierra Leone." *Social Science and Medicine* 21, no. 3 (1985): 275–82.

Boinet, Edouard. "La lèpre à Hanoi (Tonkin)." *Revue de Médecine* 10 (1890): 609–60.

Boomgaard, Peter. "Dutch Medicine in Asia, 1600–1900." In *Warm Climates and Western Medicine: The Emergence of Tropical Medicine, 1500–1900*, edited by David Arnold, 42–64. Atlanta: Rodopi, 1996.

Borst, Charlotte. *Catching Babies*. Cambridge, MA: Harvard University Press, 1995.

Boscq, J. C. "Assistance Médicale en Indochine: La leproserie de Cu-Lao-Rong." In *La dépêche coloniale illustree: La vie econonomique de nos colonies*, edited by Colonial Rédaction et Administration, 131–40. Paris: Rédaction et Administration, 1909.

Bredin, Jean-Denis. *The Affair: The Case of Alfred Dreyfus*. Translated by Jeffrey Mehlman. New York: George Braziller, Inc., 1983.

Brunschwig, Henri. *French Colonialism 1871–1914: Myths and Realities*. New York: Praeger, 1964.

Burgess, Perry. "Lepers and Leprosy." *Scientific Monthly* 42, no. 5 (1936): 396–402.

Burns, Susan L. "From 'Leper Villages' to Leprosaria: Public Health, Nationalism and the Culture of Exclusion in Japan." In *Isolation: Places and Practices of Exclusion*, edited by Alison Bashford and Carolyn Strange, 108–18. New York: Routledge, 2003.

Cady, John F. *The Roots of French Imperialism in Eastern Asia*. Ithaca, NY: Cornell University Press, 1954.

Calmette, Albert. "Compte rendu relatif à l'organisation et au fonctionnement de l'institut de vaccine animale crée à Saigon en 1891." Saigon: Imprimerie Coloniale, 1891..

Cambodian Government. *Journal officiel du Cambodge*. Phnom Penh: Cambodian Government, 1955.

Chakrabarty, Dipesh. "Subaltern Studies and Postcolonial Historiography." *Nepantla: Views from the South* 1, no. 1 (2000): 1–32.

Chan, Ok. "Contribution à l'étude de la therapeutique traditionnelle au pays khmer: Thèse pour le doctorat en médecine." M.D. thesis, Faculte de Médecine de Paris, 1955.

Chandler, David. "Folk Memories of the Decline of Angkor in Nineteenth-Century Cambodia: The Legend of the Leper King." *Journal of the Siam Society* 61, no. 1 (1979): 54–62.

———. *A History of Cambodia*. 3rd ed. Boulder: Westview Press, 1996.

Chevalier, Louis. *Le chólera, la première épidémie du XIX siècle*. La Roche-Sur-Yon: Imprimerie Centrale de L'Ouest, 1958.

Chhem, Rethy K. "Medicine and Culture in Ancient Cambodia." *The Singapore Biochemist* 13 (2000): 23–27.

Choulean, Ang. *Les êtres surnaturels dans la religion populaire khmère*. Paris: CEDORECK, 1986.

———. "Grossesse et accouchement au Cambodge: Aspects rituals." *ASEMI* XIII (1982): 87–109.

———. "Apports indiens à la médecine traditionnelle khmère." *Journal of the European Ayurvedic Society* 2 (1992): 101–13.

———. "De la naissance à la puberté: Rites et croyances khmers." In *Enfants et sociétés d'Asie du Sud-Est*, edited by Jeannine Koubi and Josiane Massard-Vincent, 153–65. Paris: Harmattan, 1994.

Clavel, Dr. *L'Assistance Médicale indigène en Indo-Chine: Organisation et fonctionnement*. Paris: Augustin Challamel, 1908.

Coedès, George. "L'Assistance Médicale au Cambodge à la fin du XIIe siecle." *Revue médicale française d'Extrême Orient* 19, no. 1 (1941): 405–15.

———. *The Indianized States of Southeast Asia*. Honolulu: University of Hawaii Press, 1971.

Cohen, William B. *Rulers of Empire: The French Colonial Service in Africa*. Stanford, CA: Hoover Institution Press, Stanford University, 1971.

Coleman, William. *Yellow Fever in the North: The Methods of Early Epidemiology*. Madison: University of Wisconsin Press, 1987.

Colgrove, James. "Science in a Democracy": The Contested Status of Vaccination in the Progressive Era and the 1920s." *Isis* 96, no. 2 (2005): 167–91.

Condominas, G. *L'espace social à propos de l'Asie du Sud-Est*. Paris: Flammarion, 1980.

Conklin, Alice. *A Mission to Civilize: The Republican Idea of Empire in France and West Africa, 1895–1930*. Stanford, CA: Stanford University Press, 1997.

Corbin, Alain. *The Foul and the Fragrant: Odor and the French Social Imagination*. Cambridge, MA: Harvard University Press, 1986.

———. *Women for Hire: Prostitution and Sexuality in France after 1850*. Translated by Alan Sheridan. Cambridge, MA: Harvard University Press, 1990.

Cunningham, Andrew, and Perry Williams. "De-Centring The "Big Picture": The Origins of Modern Science and the Modern Origins of Science." In *The Scientific Revolution: The Essential Readings*, edited by Marcus Hellyer, 219–46. Malden, MA: Blackwell, 2003.

Curtin, Philip D. *Death by Migration: Europe's Encounter with the Tropical World in the Nineteenth Century*. Cambridge: Cambridge University Press, 1995.

Daston, Lorraine. *Things That Talk: Object Lessons from Art and Science*. New York: Zone Books, 2004.

Daughton, J. P. *An Empire Divided: Religion, Republicanism, and the Making of French Colonialism, 1880–1914*. Oxford: Oxford University Press, 2006.

Davin, Anna. "Imperialism and Motherhood." In *Tensions of Empire: Colonial Cultures in a Bourgeois World*, edited by Ann Laura Stoler and Frederick Cooper, 87–151. Berkeley: University of California Press, 1997.

De Bevoise, Ken. *Agents of Apocalypse: Epidemic Disease in the Colonial Philippines*. Princeton, NJ: Princeton University Press, 1995.

De Carné, Louis. *Travels on the Mekong: Cambodia, Laos, and Yunnan*. 2 ed. Bangkok: White Lotus Press, 2000 (1872).

Declerq, Eugene, Raymond De Vries, Kirsi Viisainen, Helga Salvesan, and Sirpa Wren. "Where to Give Birth." In *Birth by Design: Pregnancy, Maternity Care, and Midwifery in North America and Europe*, edited by Raymond De Vries. New York: Routledge, 2001.

Dedet, Jean-Pierre. *Les Instituts Pasteur d'outre-mer: Cent vingt ans de microbiologie française dans le monde*. Paris: Harmattan, 2000.

Delvert, Jean. *Le paysan cambodgien*. Paris: Mouton, 1961.

Douglas, Mary. "Sacred Contagion." In *Reading Leviticus: A Conversation with Mary Douglas*, 86–106. Sheffield, UK: Sheffield Academic Press, 1996.

Douk, Phana. "Contribution à l'étude des plantes médicinales du Cambodge." PhD thesis, Faculté de Médecine de Paris, 1965.

Dubos, René, and Jean Dubos. *The White Plague: Tuberculosis, Man, Society*. Boston: Little, Brown and Company, 1952.

Dubos, René. *Louis Pasteur, Free Lance of Science*. 2nd ed. New York: Da Capo Press, 1960.

Durbach, Nadja. "Class, Gender, and the Conscientious Objector to Vaccination, 1898–1907." *The Journal of British Studies* 41, no. 1 (2002): 58–83.

Ebihara, May. "Svay, a Rice Growing Village: 1958–1960." PhD thesis, Columbia University, 1971.

Echenberg, Myron. "Pestis Redux: The Initial Years of the Third Bubonic Plague Pandemic, 1894–1901." *Journal of World History* 13, no. 2 (2002): 429–49.

Edwards, Penny. "'Propagender': Marianne, Joan of Arc and the Export of French Gender Ideology to Colonial Cambodia (1863–1954)." In *Promoting the Colonial Idea: Propaganda and Visions of Empire in France,* edited by Tony Chafer and Amanda Sackur, 116–30. New York: Palgrave Publishers, 2002.

———. *Cambodge: The Cultivation of a Nation, 1860–1945.* Honolulu: University of Hawaii Press, 2006.

Executive Committee of the Eighth Congress of the Far Eastern Association of Tropical Medicine. *Siam in 1930: General and Medical Features.* 2nd ed. Bangkok: White Lotus Press, 2000 (1930).

Faraut, F. G. *Astronomie cambodgienne.* Phnom Penh, Saigon: Imprimerie F. H. Schneider, 1910.

Fauci, Anthony. "Smallpox Vaccination Policy: The Need for Dialogue." *New England Journal of Medicine* 346, no. 17 (2002): 1319–20.

Faure, Oliver. *Les français et leur médecine aux XIXe siècle.* Paris: Belin, 1993.

Forest, Alain. *Le culte des genies protecteurs au Cambodge: Analysis et traduction d'un corpus de textes sur les Neak Ta.* Paris: Harmattan, 1992.

Foucault, Michel. *Madness and Civilization: A History of Insanity in the Age of Reason.* Translated by Richard Howard. New York: Vintage Books ed., 1988 (1965).

———. *The Birth of the Clinic.* New York: Random House, 1994.

Frieson, Kate. "Sentimental Education: Les Sages Femmes and Colonial Cambodia." *Journal of Colonialism and Colonial History* 1, no. 1 (2000): 1–27.

Furnivall, John Sydenham. *Colonial Policy and Practice: A Comparative Study of Burma and Netherlands India.* New York: New York University Press, 1956.

Furnivall, John Sydenham. *Netherlands India.* Cambridge: Cambridge University Press, 1944.

Gaide, Lucien, and Henry Bodet. "La prévention et le traitement de la lèpre en Indochine." In *Exposition coloniale internationale,* edited by Indochine Française. Section des Services d'Intérêt Social. Inspection Générale des Services Sanitaires et Médicaux de l'Indochine, 1–47. Hanoi: Imprimerie d'Extrême Orient, 1931.

———. *L'Assistance Médicale et la protection de la santé publique.* Edited by Indochine Française, Section des Services d'Intérêt Social. Inspection Générale des Services Sanitaires et Médicaux de l'Indochine, Exposition Coloniale Internationale. Hanoi: Imprimerie d'Extrême Orient, 1931.

Garnier, Francis. *Travels in Cambodia and Part of Laos: The Mekong Exploration Commission Report (1866–1868),* vol. 1. Translated by Walter E. J. Tips. Bangkok: White Lotus Press, 1996 (1885).

Geertz, Clifford. *Negara: The Theatre State in Nineteenth-Century Bali.* Princeton, NJ: Princeton University Press, 1980.

———. *The Interpretation of Culture.* New York: Basic Books, 1973.

Geison, Gerald L. *The Private Science of Louis Pasteur.* Princeton, NJ: Princeton University Press, 1995.

Gellner, Ernest. *Nations and Nationalism.* Ithaca, NY: Cornell University Press, 1983.

Good, Byron J. *Medicine, Rationality, and Experience: An Anthropological Perspective.* Edited by Anthony T. Carter, The Lewis Henry Morgan Lectures: 1990. Cambridge: Cambridge University Press, 1994.

Goscha, Christopher. *Vietnam or Indochina? Contesting Concepts of Space in Vietnamese Nationalism, 1887–1954.* Copenhagen: NIAS Books, 1995.

Grandjou, Abbé. *La Congrégation des Soeurs De La Providence De Portieux (Vosges) fondée par le vénérable Jean Martin Moyë en 1762.* Lille: Desclée, de Brouwer & Co., 1923.

Groslier, Bernard Philippe. *Angkor et le Cambodge au XVIe siècle d'après les sources portugaises et espagnoles*. Paris: Presses Universitaires de France, 1958.

Guénel, Annick. "La lutte antivariolique en Extrême-Orient: Ruptures et continuité." In *L'aventure de la vaccination*, edited by Anne Marie Moulin, 82–94. Paris: Fayard, 1996.

———. "The Creation of the First Overseas Pasteur Institute, or the Beginning of Albert Calmette's Pastorian Career." *Medical History* 43 (1999): 1–25.

Guillaume, Pierre. "Les métis en Indochine." *Annales de démographie historique* (1995): 185–95.

Guillou, Anne Yvonne. "Les médecins au Cambodge: Entre élite sociale traditionnelle et groupe professionel moderne sous influence étrangère." PhD thesis, École des Hautes Etudes en Sciences Sociales, 2001.

Gussow, Zachary. *Leprosy, Racism, and Public Health: Social Policy in Chronic Disease Control*. Boulder: Westview Press, 1989.

Ha, Marie-Paule. "Engendering French Colonial History: The Case of Indochina." In *The French and the Pacific World, 17th–19th Centuries: Explorations, Migrations, and Cultural Exchanges*, edited by Annick Foucrier, 165–95. Burlington, VT: Ashgate, 2005.

Haggis, Jane. "'Good Wives and Mothers' or 'Dedicated Workers'?" In *Maternities and Modernities: Colonial and Postcolonial Experiences in Asia and the Pacific*, edited by Margaret Jolly and Kalpana Ram, 81–93. Cambridge: Cambridge University Press, 1998.

Hammonds, Evelynn Maxine. *Childhood's Deadly Scourge: The Campaign to Control Diphtheria in New York City 1880–1930*. Baltimore: Johns Hopkins University Press, 1999.

Hankin, E. H. *Cholera in Indian Cantonments and How to Deal with It*. Allahabad: Pioneer Press, 1895.

Hansen, Anne Ruth. *How to Behave: Buddhism and Modernity in Colonial Cambodia, 1860–1930*. Honolulu: University of Hawaii Press, 2007.

Hardy, Anne. *The Epidemic Streets: Infectious Disease and the Rise of Preventive Medicine, 1856–1900*. Oxford: Clarendon Press, 1993.

Harrison, Mark. "The Tender Frame of Man: Disease, Climate, and Racial Difference in India and the West Indies, 1760–1860." *Bulletin of the History of Medicine* 70, no. 1 (1996): 68–93.

———. *Public Health in British India: Anglo-Indian Preventive Medicine 1859–1914*. Cambridge: Cambridge University Press, 1994.

Headrick, Daniel R. *The Tools of Empire: Technology and European Imperialism in the Nineteenth Century*. New York: Oxford University Press, 1981.

———. *The Tentacles of Progress: Technology Transfer in the Age of Imperialism, 1850–1940*. New York: Oxford University Press, 1988.

Headrick, Rita, and Daniel R. Headrick. *Colonialism, Health and Illness in French Equatorial Africa, 1885–1935*. Atlanta: African Studies Association Press, 1994.

Heffernan, Michael J. "A State Scholarship: The Political Geography of French International Science during the Nineteenth Century." *Transactions of the Institute of British Geographers* 19, no. 1 (1994): 21–45.

Heine-Geldern, Robert. "Conceptions of State and Kingship in Southeast Asia." *The Far Eastern Quarterly* 2, no. 1 (1942): 15–30.

Higham, Charles. *The Archaeology of Mainland Southeast Asia*. Cambridge: Cambridge University Press, 1989.

Hildreth, Martha L. "Doctors and Families in France, 1880–1930: The Cultural Reconstruction of Medicine." In *French Medical Culture in the Nineteenth Century*, edited by Ann La Berge and Mordechai Feingold, 189–209. Atlanta: Editions Rodopi, 1994.

Hoang, Bao Chau. "L'alliance des deux médecines traditionnelle et moderne au Vietnam." *Etudes vietnamiennes*, special issue on la médecine traditionelle, no. 50 (1977): 7–29.

Hou, Ngo. "Les débuts de l'Assistance Médicale au Cambodge de 1863 à 1908." M.D. thesis, University of Hanoi, 1953.

Hunt, Nancy Rose. "Le Bébé en Brousse: European Women, African Birth Spacing, and Colonial Intervention in Breast Feeding in the Belgian Congo." In *Tensions of Empire: Colonial Cultures in a Bourgeois World*, edited by Ann Laura Stoler and Frederick Cooper, 287–321. Berkeley: University of California Press, 1997.

Jacques, Claude. "Les édits des hopitaux de Jayavarman VII." *Études cambodgiennes*, no. 13 (1968): 14–17.

Jeanselme, Edouard. *Étude sur la lèpre dans la péninsule indo-chinoise et dans le Yunnan*. Paris: Carré et Naud, 1900.

Jennings, Eric Thomas. *Vichy in the Tropics: Pétain's National Revolution in Madagascar, Guadeloupe, and Indochina, 1940–1944*. Stanford, CA: Stanford University Press, 2001.

Jolly, Margaret. "Colonial and Postcolonial Plots in the History of Maternities and Modernities." In *Maternities and Modernities: Colonial and Postcolonial Experiences in Asia and the Pacific*, edited by Margaret Jolly and Kalpana Ram, 1–25. Cambridge: Cambridge University Press, 1998.

Kakar, Sanjiv. "Leprosy in British India, 1860–1940: Colonial Politics and Missionary Medicine." *Medical History* 40 (1996): 215–30.

Katz, Jonathan Glustrom. *Murder in Marrakesh: Émile Mauchamp and the French Colonial Adventure*. Bloomington: Indiana University Press, 2006.

Kermorgant, A. "Maladies epidémiques et contagieuses qui ont régné dans les colonies françaises en 1901." *Annales d'hygiène et de médecine coloniales* 6 (1903): 605–35.

———. "Notes sur la lèpre dans nos diverses possessions coloniales." *Annales d'hygiène et de médecine coloniales* 8 (1905): 25–67.

Kevles, Daniel J. *In the Name of Eugenics: Genetics and the Uses of Human Heredity*. 2nd ed. Cambridge, MA: Harvard University Press, 1995.

Khin, Sok. *L'annexion du Cambodge par les vietnamiens au XIX siècle*. Paris: Editions You-Feng, 2002.

Kipp, Rita Smith. "The Evangelical Uses of Leprosy." *Social Science & Medicine* 39, no. 2 (1994): 165–78.

Klaus, Alisa. *Every Child a Lion: The Origins of Maternal and Infant Health Policy in the United States and France, 1890–1920*. Ithaca, NY: Cornell University Press, 1993.

Knecht-van Eekelen, A. de, G. M. van Heteren, M. J. D. Poulissen, and A. M. Luyendijk-Elshout, eds. *Dutch Medicine in the Malay Archipelago 1816–1942*. Amsterdam: Rodopi, 1989.

Koven, Seth, and Sonya Michel. "Womanly Duties: Maternalist Politics and the Origins of Welfare States in France, Germany, Great Britain, and the United States, 1880–1920." *American Historical Review* 95, no. 4 (1990).

Kudlick, Catherine Jean. *Cholera in Post-Revolutionary Paris: A Cultural History*. Berkeley: University of California Press, 1996.

La Berge, Ann F. *Mission and Method: The Early Nineteenth-Century French Public Health Movement*. Cambridge: Cambridge University Press, 1992.

La Berge, Ann, and Mordechai Feingold, eds. *French Medical Culture in the Nineteenth Century*. Atlanta: Editions Rodopi, 1994.

Lapeysonnie. *La médecine coloniale: Mythes et réalités*. Paris: Seghers, 1988.

Latour, Bruno. *The Pasteurization of France*. Translated by Alan Sheridan and John Law. Cambridge, MA: Harvard University Press, 1988.

Leclère, Adhemard. *Histoire du Cambodge depuis le premier siècle de notre ère*. Paris: Librarie Paul Geuthner, 1914.

Léonard, Jacques. *La médecin entre les pouvoirs et les savoirs*. Paris: Aubier Montaigne, 1981.

Levine, Philippa. "Rereading the 1890s: Venereal Disease as "Constitutional Crisis" in Britain and British India." *The Journal of Asian Studies* 55, no. 3 (1996): 585–612.

Lieberman, Victor. *Strange Parallels: Southeast Asia in Global Context, c. 800–1830*. Cambridge: Cambridge University Press, 2003.

Löwy, Ilana. "From Guinea Pigs to Man: The Development of Haffkine's Anticholera Vaccine." *Journal of the History of Medicine and Allied Sciences* 47, no. 3 (1992): 270–309.

Luong, Hy Van. "Agrarian Unrest from an Anthropological Perspective: The Case of Vietnam." *Comparative Politics* 17, no. 2 (1985): 153–74.

MacLeod, Roy, and Milton Lewis, eds. *Disease, Medicine, and Empire: Perspectives on Western Medicine and the Experience of European Expansion*. New York: Routledge, 1988.

MacLeod, Roy. "On Visiting the 'Moving Metropolis': Reflections on the Architecture of Imperial Science." In *Scientific Aspects of European Expansion*, edited by William K. Storey, 22–55. Brookfield, VT: Variorum, 1996.

Manderson, Lenore. "Roasting, Smoking and Dieting in Response to Birth: Malay Confinement in Cross-Cultural Perspective." *Social Science & Medicine* 15B, no. 4 (1981): 509–20.

———. *Sickness and the State: Health and Illness in Colonial Malaya, 1870–1940*. Hong Kong: Cambridge University Press, 1996.

Marks, Harry M. *The Progress of Experiment: Science and Therapeutic Reform in the United States, 1900–1990*. Cambridge: Cambridge University Press, 1997.

Martel, Gabrielle. *Lovea, village des environs d'Angkor: Aspects démographiques, economiques et sociologiques du monde rural cambodgien dans la province de Siem-Reap*. Paris: École Française d'Extrême Orient, 1975.

Martine, Marie Alexandrine. "Elements de médicine traditionnelle khmer." *Seksa Khmer* 1, no. 6 (1983): 135–70.

Matras-Troubetzkoy, Jacqueline. *Un village en forêt: L'essartage chez les Brou du Cambodge*. Paris: SELAF, 1983.

Maurel, E. "Mémoire sur l'anthropologie des divers peuples vivant actuellement au Cambodge." *Mémoires de la Société d'Anthropologie de Paris* Series 2, no. T.4 (1893): 459–535.

May, Jacques. *Un médecin français en Extrême-Orient*. Paris: Editions de la Paix, 1951.

Mazet, J. *La condition juridique des métis dans les possessions françaises*. Paris: Editions Domat-Montchrestien, 1932.

McClellan, James. "Science, Medicine and French Colonialism in Old-Regime Haiti." In *Science, Medicine, and Cultural Imperialism*. New York: St. Martin's Press, 1991.

McKeown, Thomas. *The Modern Rise of Population*. New York: Academic Press, 1976.

Meade, Teresa, and Mark Walker, eds. *Science, Medicine, and Cultural Imperialism*. New York: St. Martin's Press, 1991.

Menaut, Bernard, and H. Baisez. *La lèpre au Cambodge*. Hanoi: Imprimerie d'Extrême-Orient, 1919.

Menaut, Bernard. "Matière médicale cambodgienne." In *Exposition Coloniale Internationale*, edited by Inspection Générale des Services Sanitaires et Médicaux de l'Indochine. Hanoi: Imprimerie d'Extrême-Orient, 1931.

Meyer, Charles. "Lepers in Troeung." *Études cambodgiennes* 13 (1968).

Mollaret, Henri H., and Jacqueline Brossollet. *Alexandre Yersin: 1863–1943, Un pasteurien en Indochine*. Paris: Belin, 1993.

Monnais-Rousselot, Laurence. *Médecine et colonisation: L'aventure indochinoise*. Paris: CNRS Editions, 1999.

Morizon, Réné. *Monographie du Cambodge*, Exposition Coloniale Internationale, Paris. Hanoi: Imprimerie d'Extrême-Orient, 1931.

Moseley, Jack. "Travels of Alexandre Yersin: Letters of a Pastorian in Indochina, 1890–1894." *Perspectives in Biology and Medicine* (1981): 607–18.

Moulin, Anne Marie. "Bacteriological Research and Medical Practice." In *French Medical Culture in the Nineteenth Century*, edited by Ann La Berge and Mordechai Feingold, 327–49. Atlanta: Editions Rodopi, 1994.

Muller, Gregor. *Colonial Cambodia's "Bad Frenchmen": The Rise of French Rule and the Life of Thomas Caraman, 1840–1887*. New York: Routledge, 2006.

Naono, Atsuko. "The State of Vaccination: British Doctors, Indigenous Cooperation, and the Fight against Smallpox in Colonial Burma." Ph.D. thesis, University of Michigan, 2005.

Navon, Liora. "Beggars, Metaphors, and Stigma: A Missing Link in the Social History of Leprosy." *Social History of Medicine* 11, no. 1 (1998): 89–105.

Nogue, J. "Missions de vaccine au Cambodge." *Annales d'hygiène et de médecine coloniales* (1898): 169–234.

Norodom, Ritharasi. "L'évolution de la médecine au Cambodge." M.D. thesis, Librarie Louis Arnett, 1929.

Nye, Robert A. *Crime, Madness, and Politics in Modern France: The Medical Concept of National Decline*. Princeton, NJ: Princeton University Press, 1984.

Oakley, Ann. *The Captured Womb: A History of the Medical Care of Pregnant Women*. New York: Blackwell Publishers, 1984.

Obregón, Diana. "Building National Medicine: Leprosy and Power in Colombia, 1870–1910." *Social History of Medicine* 15, no. 1 (2002): 89–108.

Ong, Aihwa. "The Production of Possession: Spirits and the Multinational Corporation in Malaysia." In *Women, Gender, Religion: A Reader*, edited by Elizabeth Castelli, 345–65. New York: Palgrave Press, 2001.

Osborne, Milton E. *The French Presence in Cochinchina: Rule and Response (1859–1905)*. Ithaca, NY: Cornell University Press, 1969.

———. "Continuity and Motivation in the Vietnamese Revolution: New Light from the 1930's." *Pacific Affairs* 47, no. 1 (1974): 37–55.

Oudshoorn, Nelly, and T. J. Pinch. *How Users Matter: The Co-Construction of Users and Technologies*. Cambridge, MA: MIT Press, 2003.

Pannetier, Dr. Adrien. *Au coeur du pays khmer: Notes cambodgiennes*. 2nd ed. Paris: CEDORECK, 1983.

Panzac, Daniel. *Quarantaines et lazarets: L'Europe et la peste d'Orient*. Aix-en-Provence: Édisud, 1986.

Parascandola, John. "Miracle at Carville: The Introduction of the Sulfones for the Treatment of Leprosy." *Pharmacy in History* 40, no. 2 (1998): 59–66.

Pati, Biswamoy, and Mark Harrison, eds. *Health, Medicine, and Empire: Perspectives on Colonial India*. Hyderabad: Orient Longman Ltd., 2001.

Patterson, K. David. *Health in Colonial Ghana: Disease, Medicine, and Socio-Economic Change, 1900–1955*. Waltham, MA: Crossroads Press, 1981.

Paul, B. D., and W. J. Demarest. "Citizen Participation Overplanned: The Case of a Health

Project in the Guatemalan Community of San Pedro La Laguna." *Social Science and Medicine* 19, no. 3 (1984): 185–92.

Paul, Harry W. *From Knowledge to Power: The Rise of the Science Empire in France, 1860–1939*. Cambridge: Cambridge University Press, 1985.

Pegg, Mark Gregory. "Le corps et l'autorité: La lèpre de Baudoin IV." *Annales: Economies sociétés civilisations* 45, no. 2 (1990): 265–88.

Pfaffenberger, Bryan. "Social Anthropology of Technology." *Annual Review of Anthropology* 21 (1992): 491–516.

Piat, Martin. "Médecine populaire au Cambodge." *Bulletin de la Sociéte des Études Indochinoises* XL, no. 4 (1965): 1–15.

Pigg, Stacy Leigh "Found in Most Traditional Societies': Traditional Medical Practitioners between Culture and Development." In *International Development and the Social Sciences*, edited by Frederick Cooper and Randall Packard, 259–90. Berkeley: University of California Press, 1997.

———. "Acronyms and Effacement: Traditional Medical Practitioners (TMP) in International Health Development " *Social Science & Medicine* 41, no. 1 (1995): 47–68.

Pluchon, Pierre. *Histoire des médecins et pharmaciens de marine et des colonies*. Toulouse: Editions Privat, 1985.

Ponchaud, François. *La cathédrale de la rizière: 450 ans d'histoire de l'eglise au Cambodge*. Paris: Librarie Arthème Fayard, 1990.

Porée-Maspero, Evelyn. "La cérémonie de l'appel des esprits vitaux chez les cambodgiens." *BEFEO* XLV, no. 1 (1951): 145–83.

———. *Cérémonies privées des cambodgiens*. Phnom Penh: Éditions de L'Institut Bouddhique, 1958.

———. "Notes sur les particularités du culte chez les cambodgiens." *BEFEO* 54, no. 2 (1950): 619–41.

Prakash, Gyan. *Another Reason: Science and the Imagination of Modern India*. Princeton, NJ: Princeton University Press, 1999.

Primet. "Rapport sur la prophylaxie de la lèpre en Indochine." Hanoi-Haiphong: Gouvernement Général de l'Indochine, Direction Générale de la Santé, 1910.

Pyenson, Lewis. *Civilizing Mission: Exact Sciences and French Overseas Expansion, 1830–1940*. Baltimore: Johns Hopkins University Press, 1993.

———. *Empire of Reason: Exact Sciences in Indonesia, 1840–1940*. New York: E. J. Brill, 1989.

Pym, Christopher, ed. *Henri Mouhot's Diary: Travels in the Central Parts of Siam, Cambodia, and Laos During the Years 1858–61*, Oxford in Asia Historical Reprints. Kuala Lumpur: Oxford University Press, 1966.

Rabinow, Paul. *French Modern*. Cambridge, MA: MIT Press, 1989.

Raillon, Louis, and Madeline Raillon. *Jean Cassaigne, la lèpre et Dieu*. Paris: Editions Saint-Paul, 1993.

Reid, Anthony. *Southeast Asia in the Age of Commerce: 1450–1680, Volume One: The Lands Below the Winds*. Vol. 1. New Haven: Yale University Press, 1988.

Richell, Judith. "Ephemeral Lives: The Unremitting Infant Mortality of Colonial Burma, 1891–1941." In *Women and Children First: International Maternal and Infant Welfare, 1870–1945*, edited by Valerie Fildes, Lara Marks, and Hilary Marland, 133–53. New York: Routledge, 1992.

Robequain, Charles. *The Economic Development of French Indo-China*. New York: Oxford University Press, 1944.

Rosenberg, Charles E. *The Care of Strangers: The Rise of America's Hospital System*. New York: Basic Books, Inc., 1987.

———. *The Cholera Years*. Chicago: University of Chicago Press, 1962.

Rosenkrantz, Barbara. *Public Health and the State: Changing Views in Massachusetts, 1842–1936*. Cambridge, MA: Harvard University Press, 1972.

San Antonio, Father Gabriel Quiroga de. *A Brief and Truthful Relation of Events in the Kingdom of Cambodia*. Translated by Antoine Cabaton. Bangkok: White Lotus Press, 1998 (1604).

Sargent, Carolyn, and John Marcucci. "Aspects of Khmer Medicine among Refugees in Urban America." *Medical Anthropology Quarterly*, new series 16, no. 1 (1984): 7–9.

Scheper-Hughes, Nancy, and Margaret Lock. "Mindful Body: A Prolegomenon to Future Work in Medical Anthropology." *Medical Anthropology Quarterly*, new series 1, no. 1 (1987): 6–41.

Schneider, William. *Quality and Quantity: The Quest for Biological Regeneration in Twentieth-Century France*. New York: Cambridge University Press, 1990.

Scott, James C. *The Moral Economy of the Peasant: Rebellion and Subsistence in Southeast Asia*. New Haven: Yale University Press, 1976.

Seng, Loh Kah. "'Our Lives Are Bad But Our Luck Is Good': A Social History of Leprosy in Singapore." *Social History of Medicine* 21, no. 2 (2008): 291–309.

Silla, Eric. *People Are Not the Same: Leprosy and Social Identity in Twentieth-Century Mali*. Portsmouth, NH: Heinemann, 1998.

Sorensen, W., P. B. Anderson, R. Speaker, and J. E. Vilches. "Assessment of Condom Use among Bolivian Truck Drivers through the Lens of Social Cognitive Theory." *Health Promotion International* 22, no. 1 (2007): 37–43.

Souk-Aloun, Phou Ngeun. *La médecine bouddhique traditionnelle en pays théravâda*. Limoges: Editions Roger Jollois, 1995.

Souyris-Rolland, André. "Les procédés magiques d'immunisation chez les cambodgiens." *Bulletin de la Sociéte des Etudes Indochinoises* XXVI, no. 2 (1951): 175–87.

Starr, Paul. *The Social Transformation of American Medicine*. New York: Basic Books, Inc., 1982.

Stockwell, A. J. "Southeast Asia in War and Peace: The End of European Colonial Empires." In *The Cambridge History of Southeast Asia: Volume 2 Part Two: From World War II to the Present*, edited by Nicholas Tarling, 329–86. Cambridge: Cambridge University Press, 1992.

Stoler, Ann Laura. "Rethinking Colonial Categories: European Communities and the Boundaries of Rule." In *Colonialism and Culture*, edited by Nicholas B. Dirks, 319–52. Ann Arbor: University of Michigan Press, 1992.

———. "Sexual Affronts and Racial Frontiers." In *Tensions of Empire: Colonial Cultures in a Bourgeois World*, edited by Ann Laura Stoler and Frederick Cooper, 198–237. Berkeley: University of California Press, 1997.

Sullivan, Norma. *Masters and Managers: A Study of Gender Relations in Urban Java*, Women in Asia Publication Series. St. Leonards, Australia: Allen & Unwin, 1994.

Sullivan, Rodney. "Cholera and Colonialism in the Philippines, 1899–1903." In *Disease, Medicine, and Empire: Perspectives on Western Medicine and the Experience of European Expansion*, edited by Roy MacLeod and Milton Lewis, 184–300. New York: Routledge, 1988.

Ta-Kouan, Tcheou. *Mémoires sur les coutumes du Cambodge*. Translated by Paul Pelliot. Paris: Librarie d'Amérique et d'Orient, 1951.

Tan, Boun Suy. "Introduction du quinquina au Cambodge." *Seksa Khmer* 1, no. 3 (1981): 49–58.

Tana, Li. *Nguyen Cochinchina: Southern Vietnam in the Seventeenth and Eighteenth Centuries.* Ithaca, NY: Southeast Asia Program Publications, 1998.

Tarling, Nicholas, ed. *The Cambridge History of Southeast Asia: Volume 2 Part One: From c. 1800 to the 1930s*. Vol. 2. Cambridge: Cambridge University Press, 1992.

Tauch, Chhuong. *Battambang during the Time of the Lord Governor*. Phnom Penh: CEDORECK, 1994.

Taylor, Jean Gelman. *The Social World of Batavia: European and Eurasian in Dutch Asia*. Madison: University of Wisconsin Press, 1983.

Thierry, Solange. *Le Cambodge des contes*. Paris: Editions L'Harmattan, 1985.

Thiroux. "Contribution à l'étude de la contagion et de la pathogénie de la lèpre." *Annales d'hygiène et de médecine coloniales* 6 (1903): 564–82.

Thomas, Martin. *The French Empire at War, 1940–45*. Manchester, UK: Manchester University Press, 1998.

Thomas, Nicholas. *Entangled Objects: Exchange, Material Culture, and Colonialism in the Pacific*. Cambridge, MA: Harvard University Press, 1991.

Tomes, Nancy. *The Gospel of Germs: Men, Women and the Microbe in American Life*. Cambridge, MA: Harvard University Press, 1998.

Tully, John. *France on the Mekong: A History of the Protectorate in Cambodia, 1863–1953*. Lanham, MD: University of America Press, 2002.

———. *Cambodia under the Tricolour: King Sisowath and the 'Mission Civilisatrice' 1904–1927*. Clayton, Victoria, Australia: Monash Asia Institute, 1996.

Turner, Brian S. *Medical Power and Social Knowledge*. 2nd ed. London: Sage Publications, 1995.

Unschuld, Paul. *Medicine in China: A History of Ideas*. Berkeley: University of California Press, 1985.

Van der Geest, Sjaak, Susan Reynolds Whyte, and Anita Hardon. "The Anthropology of Pharmaceuticals: A Biographical Approach." *Annual Review of Anthropology* 25 (1996): 153–78.

Van Esterik, Penny, ed. *Women of Southeast Asia*. DeKalb, IL: Northern Illinois University, Center for Southeast Asian Studies, 1982.

Vaudon, Jean. "Les Filles de Saint-Paul de Chartes. Une léproserie en Indochine." *Revue d'histoire des missions* 3, no. 4 (1926): 541–48.

Vaughan, Megan. *Curing Their Ills: Colonial Power and African Illness*. Stanford, CA: Stanford University Press, 1991.

Warner, John Harley. *The Therapeutic Perspective: Medical Practice, Knowledge, and Identity in America, 1820–1885*. Princeton, NJ: Princeton University Press, 1997.

Watson, C. W., and Roy Ellen, ed. *Understanding Witchcraft and Sorcery in Southeast Asia*. Honolulu: University of Hawaii Press, 1993.

Weber, Eugen. *The Hollow Years: France in the 1930s*. New York: W. W. Norton, 1996.

Weber, Max. *The Theory of Social and Economic Organization*. New York: The Free Press, 1947.

Weindling, Paul. "The Origins of Informed Consent: The International Scientific Commission on Medical War Crimes, and the Nuremburg Code." *Bulletin of the History of Medicine* 75 (2001): 37–71.

Weisz, George. "The Politics of Medical Professionalization in France 1845–1848." *Journal of Social History* 12, no. 1 (1978): 3–30.

White, Patrice M. "Crossing the River: Traditional Beliefs and Practices of Khmer Women during Pregnancy, Birth and Postpartum." 67. Phnom Penh: PACT/JSI and Cambodian Ministry of Health, 1995.

Williams, Naomi. "The Implementation of Compulsory Health Legislation: Infant Smallpox Vaccination in England and Wales, 1840–1890." *Journal of Historical Geography* 20 (1994): 396–412.

Witz, Anne. "'Colonising Women': Female Medical Practice in Colonial India, 1880–1890." In *Women and Modern Medicine,* edited by Lawrence Conrad and Anne Hardy, 23–52. Amsterdam: Editions Rodopi B.V., 2001.

Wolters, O. W. *History, Culture and Region in the Southeast Asian Perspectives.* Pasir Panjang, Singapore: Institute of Southeast Asian Studies, 1982.

Wright, Gordon. *France in Modern Times.* 5th ed. New York: W. W. Norton, 1995.

Wright, Gwendolyn. *The Politics of Design in French Colonial Urbanism.* Chicago: University of Chicago Press, 1991.

Zinoman, Peter. *The Colonial Bastille: A History of Imprisonment in Vietnam, 1862–1940.* Berkeley: University of California Press, 2001.

# INDEX

The letter *f* following a page number denotes a figure.